Anne Kress

Stable isotope dendroclimatology in the Swiss Alps:

Anne Kress

Stable isotope dendroclimatology in the Swiss Alps:

A 1200-year record from European larch

Südwestdeutscher Verlag für Hochschulschriften

Impressum/Imprint (nur für Deutschland/ only for Germany)
Bibliografische Information der Deutschen Nationalbibliothek: Die Deutsche Nationalbibliothek verzeichnet diese Publikation in der Deutschen Nationalbibliografie; detaillierte bibliografische Daten sind im Internet über http://dnb.d-nb.de abrufbar.
Alle in diesem Buch genannten Marken und Produktnamen unterliegen warenzeichen-, marken- oder patentrechtlichem Schutz bzw. sind Warenzeichen oder eingetragene Warenzeichen der jeweiligen Inhaber. Die Wiedergabe von Marken, Produktnamen, Gebrauchsnamen, Handelsnamen, Warenbezeichnungen u.s.w. in diesem Werk berechtigt auch ohne besondere Kennzeichnung nicht zu der Annahme, dass solche Namen im Sinne der Warenzeichen- und Markenschutzgesetzgebung als frei zu betrachten wären und daher von jedermann benutzt werden dürften.

Verlag: Südwestdeutscher Verlag für Hochschulschriften Aktiengesellschaft & Co. KG
Dudweiler Landstr. 99, 66123 Saarbrücken, Deutschland
Telefon +49 681 37 20 271-1, Telefax +49 681 37 20 271-0
Email: info@svh-verlag.de
Zugl.: Zürich, ETH, Diss., 2009

Herstellung in Deutschland:
Schaltungsdienst Lange o.H.G., Berlin
Books on Demand GmbH, Norderstedt
Reha GmbH, Saarbrücken
Amazon Distribution GmbH, Leipzig
ISBN: 978-3-8381-1812-3

Imprint (only for USA, GB)
Bibliographic information published by the Deutsche Nationalbibliothek: The Deutsche Nationalbibliothek lists this publication in the Deutsche Nationalbibliografie; detailed bibliographic data are available in the Internet at http://dnb.d-nb.de.
Any brand names and product names mentioned in this book are subject to trademark, brand or patent protection and are trademarks or registered trademarks of their respective holders. The use of brand names, product names, common names, trade names, product descriptions etc. even without a particular marking in this works is in no way to be construed to mean that such names may be regarded as unrestricted in respect of trademark and brand protection legislation and could thus be used by anyone.

Publisher: Südwestdeutscher Verlag für Hochschulschriften Aktiengesellschaft & Co. KG
Dudweiler Landstr. 99, 66123 Saarbrücken, Germany
Phone +49 681 37 20 271-1, Fax +49 681 37 20 271-0
Email: info@svh-verlag.de

Printed in the U.S.A.
Printed in the U.K. by (see last page)
ISBN: 978-3-8381-1812-3

Copyright © 2010 by the author and Südwestdeutscher Verlag für Hochschulschriften Aktiengesellschaft & Co. KG and licensors
All rights reserved. Saarbrücken 2010

There have been joys too great to be described in words, and there have been griefs upon which I have not dared to dwell; and with these in mind I say: Climb if you will, but remember that courage and strength are naught without prudence, and that a momentary negligence may destroy the happiness of a lifetime. Do nothing in haste; look well to each step; and from the beginning think what may be the end.

Edward Whymper in "Scrambles Amongst the Alps" (1865)

Photo: Bietschhorn, Switzerland, August 2008

Contents

Abbreviations	**v**
Summary	**ix**
Zusammenfassung	**xiii**

1 General introduction   **1**
    1.1 Status of paleoclimatic research 4
    1.2 Has palaeoclimate research helped? Where are we now? 9
    1.3 The Millennium project 11
    1.4 This study: objectives and research questions 15
    1.5 Thesis structure 17

2 Stable isotopes in dendroclimatology   **29**
    2.1 Stable isotope theory 31
    2.2 State-of-the-art in stable isotope dendroclimatology 39

3 Stable isotope coherence in the earlywood and latewood of tree-line conifers   **53**
    3.1 Introduction 55
    3.2 Materials and methods 57
    3.3 Results 58
    3.4 Discussion and conclusions 63

4 Summer temperature dependency of larch budmoth outbreaks   **73**
    4.1 Introduction 74

Contents

    4.2   Material and methods 77
            4.2.1   Sampling strategy 77
            4.2.2   Sample analysis 77
            4.2.3   Climatic data 80
            4.2.4   Grey larch budmoth and stable isotopes 81
    4.3   Results 82
            4.3.1   Climate-isotope relationship (AD 1900-2004) 82
            4.3.2   Tree-growth, LBW outbreaks, and climate 84
            4.3.3   Site-specific aspects of LBM dynamics (AD 1900–2004) 85
            4.3.4   Species-specific response 87
            4.3.5   Long-term climate forcing and LBM outbreaks 88
    4.4   Discussion 90

5 A 350-year drought reconstruction from Alpine tree-ring stable isotopes 103
    5.1   Introduction 104
    5.2   Methods 107
            5.2.1   Study site and sampling strategy 107
            5.2.2   Sample analysis 109
            5.2.3   Climate data 110
            5.2.4   Drought index 111
    5.3   Results 112
            5.3.1   Signal strength in carbon and oxygen isotope series 112
            5.3.2   Climate signal(s) 115
            5.3.3   Identification of the dominating climate signal in carbon isotopes 118
            5.3.4   Drought reconstruction 123
    5.4   Discussion and Conclusions 127

6 1200 years of climate history from stable carbon and oxygen isotopes 141
    6.1   Introduction 142
    6.2   Selection and processing of material 144
            6.2.1   The Lötschental 146
            6.2.2   The Simplon region 147
            6.2.3   Sampling design 147
            6.2.4   Isotope analysis 150
            6.2.5   Statistical analysis 151

    6.3   Results 151
            6.3.1   The raw carbon and oxygen chronologies 151
            6.3.2   Linking the cohorts: the standardized chronologies 155
            6.3.3   The climate signal(s) in the standardized chronologies 155
    6.4   Discussion 161
            6.4.1   Climate signals 161
            6.4.2   Strengths and remaining limitations of the standardized isotope series 164
    6.5   Conclusions 166

7 Conclusions and outlook   179

List of Tables   I

List of Figures   IV

Acknowledgements   V

Abbreviations

ABS Mean absolute difference

AD Anno Domini

AOGCM Atmosphere ocean general circulation model

BUT Slovakian scPDSI drought reconstruction

CE Coefficient of efficiency

CRU Climate Research Unit

CV Coefficient of variation

DRI Drought index

DW Durbin-Watson statistic

EPS Expressed population signal

ESB Eastern spruce budworm

ETH Eidgenössische Technische Hochschule Zürich

EU European Union

EW Earlywood

GAR Greater Alpine region

GCM General circulation model

Abbreviations

HISTALP Historical instrumental climatological surface time series of the greater Alpine region

IPCC Intergovernmental Panel on Climate Change

KNMI Koninklijk Nederlands Meteorologisch Instituut

LBM Larch budmoth (*Zeiraphera diniana* Gn.)

LIA Little Ice Age

LOE Lötschental

LOT$_{DRI}$ Lötschental carbon-isotope drought reconstruction

LW Latewood

m a.s.l. Meter above sea level

MSE Mean squared error

MTM Multi-taper method

MWP Medieval Warm Period

MXD Maximum latewood density

NAO North Atlantic Oscillation

NAS National Academy of Sciences

NH Northern Hemisphere

OBH In valley spring precipitation reconstruction

PDSI Palmer drought severity index

PSI Paul Scherrer Institut

RBAR Inter-series correlation

r Pearson's correlation coefficient

RE Reduction of error

RSME Root mean squared error

scPDSI Self-calibrated Palmer drought severity index

SEA Superposed epoch analysis

SIM Simplon

SPM Summary for Policymakers (IPCC)

SRES Special report on emission scenarios

STDEV Standard deviation

TAR Third Assessment Report (IPCC)

TRW Tree-ring width

VPDB Vienna Pee Dee Belemnite

VSMOW Vienna Standard Mean Ocean Water

WLR Bavarian Forest spring-summer precipitation reconstruction

WMS Vienna Basin summer precipitation reconstruction

WR Whole ring

WSL Eidgenössiche Forschungsanstalt für Wald, Schnee und Landschaft

WUE Water-use efficiency

Summary

There is little doubt that the Earth's climate is warming and that anthropogenic greenhouse gas emissions are chiefly to blame. However, to improve the models that are used to predict future climate change, a better assessment is needed of how climate has varied in the past, both temporally and spatially. To predict the impact of climate change on landscapes, ecosystems and the Earth's population, a broader understanding of the range of climate conditions that prevailed in the past is required. In this context, paleoclimatic research is of great importance. Many natural climate archives are spatially limited (e.g., ice cores) or lack high temporal resolution (e.g., marine sediments) and proxies recording temperature are considerably over-represented. Reconstructions of past moisture variability are rare, especially over central Europe. To overcome these restrictions and limitations, the EU-funded Millennium project (2006–2010) seeks to investigate European climate variability over the last 1,000 years using a wide range of instrumental, documentary and natural climate archives, including tree-ring stable isotopes, as they can provide a variety of climatic information (Chapter 2). This multi-proxy ensemble should improve model predictions of future climate change and therefore help to better assess its associated impacts.

As a part of the Millennium project, this thesis involved the compilation of a carbon and an oxygen tree-ring isotope chronology from European larch (*Larix decidua* Mill.) in the Swiss Alps, the currently longest available tree-ring isotope chronologies in Europe. Establishing these 1,200-year long isotope records involved the sampling of both living trees and historical timber. Sampling, sample processing and isotope measurements were accomplished in cooperation with the Swiss Federal Research Institute WSL and the University of Bern. With these two isotope chronologies, the following

Summary

objectives were addressed: (i) to identify the dominating climate signal in carbon and oxygen isotopes, (ii) to assess any potential biological biases, and (iii) to reconstruct long-term climate variability from stable isotopes.

Initially, two subsets of tree-ring stable isotopes from two European tree-line locations were compared to test the homogeneity of signals between tree-ring earlywood, latewood and whole-ring cellulose. Scots pine (*Pinus sylvestris* L.) from northern Norway and European larch (*Larix decidua* Mill.) from the Swiss Alps demonstrated a high common signal between earlywood and latewood, correlating strongly with summer temperature. These results suggest that, for European tree-line conifers, the separation of earlywood from latewood is unnecessary to resolve an annual isotopic signal and to provide accurate climate calibrations. Indeed, using the whole ring may even improve climate correlations and therefore climate reconstructions. Thus, the use of whole-ring cellulose is recommended for climate reconstructions from tree-ring stable isotopes in conifers at tree-line location (Chapter 3).

Since European larch (*Larix decidua* Mill.) in the interior valleys of the European Alps is periodically infested by the foliage-feeding larch budmoth (*Zeiraphera diniana* Gn.), possible impacts on the isotopic signatures in tree rings were investigated to exclude potential errors in the climate reconstructions. By comparing carbon and oxygen isotope chronologies of larch and their corresponding tree-ring widths from the Lötschental and the Simplon region against isotope data from non-host spruce (*Picea abies*), any effect of larch budmoth outbreaks on the isotopic signatures could be excluded. These results were confirmed by robust climate-isotope relationships that persist back in time despite several severe, well documented outbreak events in the past. A comparison with long-term monthly resolved temperature data furthermore revealed a strong coherence between late-summer temperatures and larch budmoth defoliation events, suggesting that cool summers are conducive to outbreak events, while at the same time being accurately replicated in the stable isotope signatures (Chapter 4).

A sensitivity study focussed on the climate-isotope relationships in living trees from the Lötschental. Unlike the more traditional dendroclimatological parameters tree-ring width and maximum latewood density, which contain only summer temperature information at this site, both isotope series demonstrated in addition to temperature a highly significant sensitivity to precipitation (mainly for carbon) and sunshine duration (mainly for oxygen). Although being restricted to a narrow July-August window, all of these climate-isotope relationships are preserved in younger trees from the same site and strong inter-tree correlations further emphasize the high degree of climate sensitivity. Oxygen isotopes, being strongly connected to meteoric (source) water origin, react to regional weather patterns, and therefore appear to be most strongly coupled with sunshine duration as meteorological variable. At the same time, the oxygen relationship to carbon indicated that temperature and precipitation may not co-vary back in time. The reconstruction of a climate variable that can account for potential instabilities between these meteorological variables was therefore made. The resulting drought index, combining temperature and precipitation, yields the first carbon isotope based summer drought reconstruction for the Swiss Alps (Chapter 5).

Finally, climate variability over the entire 1200-year length of the isotope chronologies was assessed. By combining different cohorts from living trees and historical buildings, inconsistent offsets became apparent, which could not be explained by methodological discrepancies during sample analysis. Thus the cohorts were standardized before they were joined to avoid any overemphasis of any non-climate related long-term trends. Frequency analyses of the resulting time series demonstrated that some of the low frequency variability was maintained in these records, indicating decadal- to centennial-scale variability for July-August moisture variability (carbon) and sunshine duration (oxygen). In particular, an 11-year periodicity was identified in the oxygen isotope record, which appears likely to reflect periodic oscillations in solar activity. In combination with an existing temperature reconstruction, the isotopes clearly divide the major climate periods of the past millennium into wet/dry and sunny/cloudy episodes. In contrast to the rather dry period of recent warming, the "Medieval Warm Period" for example, appears to have been rather wet, suggesting good conditions for agricultural production and therefore human subsistence (Chapter 6).

Overall, the tree-ring stable isotope records presented here appear to be reliable archives of climate variability over the past 1,200 years. Their unique signals, being sensitive to both moisture and sunshine duration, allow us to ascertain past changes in central Europe's hydrological cycle and may contribute to a more accurate assessment of future climate changes.

Zusammenfassung

Es besteht wenig Zweifel daran, dass die Erde sich erwärmt und anthropogene Treibhausgase eine wesentliche Ursache darstellen. Um Vorhersagen über die Klimaveränderung zu präzisieren, sollte Klimamodellen ein umfassendes Bild der räumlichen und zeitlichen Klimavariabilität in der Vergangenheit zugrunde liegen. Dieses Wissen wird benötigt, um die Auswirkungen und Folgen des Klimawandels auf die Landschaft, ihre Ökosysteme und ihre Bevölkerung besser abschätzen zu können. In diesem Zusammenhang ist die Paläoklimaforschung von grundlegender Bedeutung. Die natürlichen Klimaarchive sind häufig räumlich limitiert (z.b. bei Eisbohrkernen) oder es fehlt ihnen an hoher zeitlicher Auflösung (z.b. bei marinen Sedimenten). Darüber hinaus sind temperatur-sensitive Archive weit in der Überzahl, während Hinweise zur vergangenen Variabilität im Wasserhaushalt, insbesondere für Zentraleuropa, sehr selten sind. Das EU-finanzierte Millennium-Projekt ist daher bestrebt, die Klimavariabilität in Europa für die vergangenen 1000 Jahre genauer zu untersuchen. Mit Hilfe verschiedenster historischer, instrumenteller und natürlicher Archive sollen modellbasierte Klimavorhersagen verbessert und damit die Auswirkungen des Klimawandels genauer abgeschätzt werden können. Stabile Isotope in Jahrringen sind hierbei von besonderem Interesse, da sie verschiedene Klimasignale enthalten können (Kapitel 2).

Die in dieser Arbeit entwickelten Chronologien stabiler Kohlenstoff- und Sauerstoffisotope aus Jahrringen der Europäischen Lärche in den Schweizer Alpen stellen einen Teil des Millennium-Projektes dar und sind die derzeit längsten Isotopenchronologien in Zentraleuropa. Um die Chronologie-Gesamtlänge von 1200 Jahren zu erreichen, wurden die Proben von lebenden Bäumen mit Holz aus historischen Gebäuden

Zusammenfassung

ergänzt. Die Probennahme, die Probenaufbereitung und die Messung der Isotope erfolgten in Zusammenarbeit mit der Eidgenössischen Forschungsanstalt für Wald, Schnee und Landschaft (WSL) und der Universität Bern. Mit diesen Isotopenchronologien wurden die folgenden Forschungsziele verfolgt: (i) Die Identifikation des dominierenden Klimasignals in der Kohlenstoff- und der Sauerstoffreihe; (ii) die Abschätzung möglicher biologischer Beeinträchtigungen im Klimasignal und (iii) die Rekonstruktion von langfristigen Klimaschwankungen.

In einem ersten Schritt wurden Isotopendaten aus Jahrringen von zwei europäischen Waldgrenzstandorten untersucht, um die Homogenität des Isotopensignals im Frühholz, Spätholz und im gesamten Jahrring zu testen. Waldkiefer (*Pinus sylvestris*) aus Nordnorwegen und Europäische Lärche (*Larix decidua* Mill.) aus den Schweizer Alpen zeigten ein stark ausgeprägtes gemeinsames Signal in allen Jahrringbestandteilen, das zudem ähnlich stark mit der Sommertemperatur korreliert ist. Diese Ergebnisse legen den Schluss nahe, dass Jahrringe von europäischen Koniferen auf Waldgrenzstandorten nicht in Früh- und Spätholz unterteilt werden müssen, um ein starkes Klimasignal der Isotope zu erhalten. Da der gesamte Jahrring sogar das Klimasignal stärker wiederzugeben vermag, ist seine Verwendung in der Isotopenanalyse höchst wünschenswert (Kapitel 3).

Da die Europäische Lärche (*Larix decidua* Mill.) in regelmäßigen Intervallen vom Grauen Lärchenwickler (*Zeiraphera diniana* Gn.) heimgesucht wird, sind mögliche Auswirkungen seiner Fraßtätigkeit auf die Isotopensignatur im Jahrring untersucht worden. Der Vergleich von Isotopen-Chronologien von Lärchen aus dem Lötschental und der Region Simplon mit Isotopendaten nicht befallener Fichten (*Picea abies*) vom gleichen Standort zeigte, dass jeglicher Effekt von Lärchenwicklerereignissen auf die Isotopensignatur ausgeschlossen werden kann. Dies fand weiterhin Bestätigung, da die Beziehungen zwischen den Isotopen und dem Klima trotz zahlreicher starker Befallsjahre zeitlich unverändert blieben. In Kombination mit ihrer Jahrringbreite und einer lang zurückreichenden Temperaturrekonstruktion enthüllten die Isotope darüber hinaus eine starke Beziehung zwischen dem Auftreten des Lärchenwicklers und der Sommertemperatur. Demnach sind Befallsjahre gekoppelt mit niedrigen Sommertemperaturen (Kapitel 4).

Eine Sensitivitätsstudie setzte den Schwerpunkt auf die Identifikation des dominierenden Klimasignals in den beiden Isotopenchronologien der rezenten Bäume des Lötschentals. Im Gegensatz zu den klassischen dendroklimatologischen Größen Jahrringbreite und -dichte, die im Lötschental ein klares Temperatursignal aufweisen, zeigten beide Isotope zusätzlich ein starkes Niederschlagssignal (vor allem die Kohlenstoffisotope) und eine Affinität zur Sonnenscheindauer (insbesondere die Sauerstoffisotope). Diese Klimasignale sind auf ein enges Juli-August-Zeitfenster beschränkt und finden Replikation in Isotopenserien von jungen Bäumen am gleichen Standort. Darüber hinaus lässt eine starke Korrelation der Bäume untereinander auf eine starke Klimasensitivität in den Isotopenchronologien schließen. Die Sauerstoffisotope sind eng gekoppelt mit der Isotopensignatur im Niederschlagswasser und damit auch abhängig von Großwetterlagen (= Herkunft der Niederschläge), die durch eine charakteristische Sonnenscheindauer gekennzeichnet sein können. Gleichzeitig weist das Kohlenstoff-Sauerstoff-Verhältnis darauf hin, dass die Beziehung zwischen den meteorologischen Variablen Temperatur und Niederschlag zeitlich variieren kann. Die Rekonstruktion von Indizes, welche die Information der beiden Parameter koppeln können, ist daher von größerer Zuverlässigkeit als die von einzelnen Klimavariablen. Dies führte zur Berechnung eines Trockenheitsindex, mit dessen Hilfe erstmals eine auf Kohlenstoffisotopen basierende Sommertrockenheitsrekonstruktion für die Schweizer Alpen entstand (Kapitel 5).

Schließlich wurden die Klima-Isotop-Beziehungen auf die gesamten 1200 Jahre angewendet. Hierzu mussten die Teilstücke zunächst sinnvoll verbunden werden. Da die Diskrepanzen zwischen den einzelnen Teilstücken keine methodologische Ursache haben, wurde jedes Teilstück über seine gesamte Länge standardisiert, bevor sie dann zu einer Chronologie verbunden wurden. Frequenzanalysen zeigten, dass auf diese Weise dekadische bis hundertjährige Trends präzise wiedergegeben werden, gleichzeitig jedoch Trends über mehrere Jahrhunderte hinweg verloren gehen können. Interessanterweise zeigte die Sauerstoffchronologie eine etwa 11-jährige Periodizität, ein Zeitraum, der von der variierenden Sonnenaktivität her bekannt ist. Zusammen mit einer Temperaturrekonstruktion konnten bekannte Klimaperioden in nasse/ trockene und sonnige/ wolkige Episoden unterteilt werden. Im Gegensatz zur heutigen Erwärmung, die von vielen trockenen Episoden begleitet ist, zeigte sich die "Mittelalterliche Wärmeperiode" als eher feucht, was auf geeignete Bedingungen für die Landwirtschaft und damit auf gute Lebensbedingungen für den Menschen schließen lässt (Kapitel 6).

Die hier vorliegenden Kohlenstoff- und Sauerstoffchronologien haben bewiesen, dass zuverlässige Klimarekonstruktionen auf der Basis stabiler Isotope aus Jahrringen

Zusammenfassung

möglich sind. Darüber hinaus zeigten sie eine Sensitivität zu Klimaparametern, über deren vergangene Variabilität wenig bekannt und die daher von besonderem Interesse ist. Durch die Feuchtesensitivität der Kohlenstoffisotope und das Sonnenscheinsignal in den Sauerstoffisotopen können Veränderungen im Wasserkreislauf zurückverfolgt werden. Diese Erkenntnisse können möglicherweise dazu beitragen, Klimaveränderungen in der Zukunft besser vorherzusagen, und sie können somit für eine Anpassung und Planung richtungsweisend sein.

1

General introduction

Climate has changed over the course of Earth's history on all time scales. At present, the Earth is about to experience a dramatic climate change which is presumably caused by the rapid increase of the Earth's population along with demands for energy, mobility and prosperity and the accompanying emission of greenhouse gases (carbon-dioxide (CO_2), methane (CH_4) and nitrous oxide (N_2O)). Current global temperatures are higher than they have ever been during at least the last 500 years and probably even for more than the past 1,300 years (Jansen et al., 2007).

Some aspects of current climate change are not unusual, but others are. Before human activity could have played a role, the principal driver of past climate variability was changes in the Earth's radiation balance. There are three fundamental ways the Earth's radiation balance can change, thereby causing climate change on different time scales: changes in the incoming solar radiation (e.g., by changes in the Earth's orbit),

Chapter 1 General introduction

changes in the fraction of solar radiation that is reflected and absorbed (e.g., by changes in cloud cover), and changes in the long-wave energy radiation returned back to space (e.g., by changes in greenhouse gas concentrations) (IPCC, 2007). Changes of the incoming solar radiation cause low- to mid-frequency oscillations in climate. While on a multi-millennial scale, they are caused by regular variations in the Earth's orbit around the sun (Milankovitch cycles), decadal- to century-scale changes can be induced by alterations in solar activity (e.g., Petit et al., 1999). Within these cycles, high-frequency inter-annual to decadal fluctuations can occur. They primarily depend on ocean and atmosphere dynamics and atmospheric gas and aerosol composition (e.g., Bonan, 2002; Chylek and Lohmann, 2008). While natural changes in climate have historically influenced human settlement, socio-economic development and culture, the relationship between human beings and climate has shifted towards an ever-increasing impact of human activity on climate (Lamb, 2005). Currently the main source of human-induced climate change are emissions associated with energy use. In particular the anthropogenic production of greenhouse gases (CO_2, CH_4 and N_2O) has reached a level that is very unusual for the Quaternary (about the last two million years) and is likely to continue to increase in the future. The concentrations of CO_2 and CH_4 in the atmosphere have reached – with an exceptionally fast rate – a record high relative to more than the past millennium (IPCC, 2007). Predictions of how these changes will affect climate are rather uncertain, mainly because of uncertainties in the sensitivity of the climate system to change but also because of imperfect knowledge of the terrestrial carbon cycle and because of the open question of how much CO_2 will be taken up by the ocean (Edwards et al., 2007).

While the globally well-mixed CO_2 concentrations can be reconstructed reliably from antarctic ice cores for the last 650,000 years, temperature is a more difficult variable to reconstruct. As temperature does not have the same value all over the globe, a single record (e.g., an ice core) is only of limited value. Nevertheless, a warming of the climate is unequivocal, as an increase of global average air and ocean temperatures, a widespread melting of snow and ice and a rising global average sea level are now evident from observations (IPCC, 2007). However, the rate and magnitude of the future warming trend remain uncertain. In Figure 1.1 the projected surface temperature is modeled for the early and late 21st century on the basis of the special report on emission scenarios (SRES)[1]. The spatial distribution of projected surface temperature

---

[1]SRES are emission scenarios developed by Nakicenovic and Swart (2000) (in IPCC, 2007) and used,

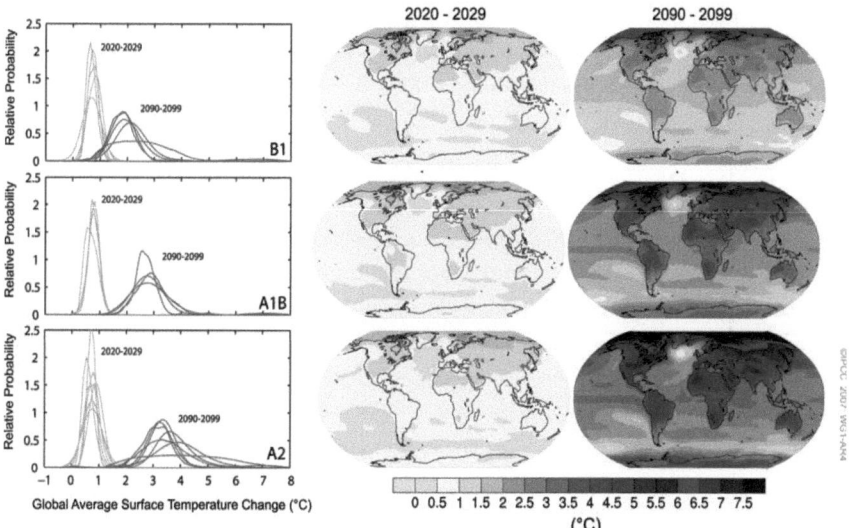

Figure 1.1: Projected surface temperature changes for the early and late 21st century relative to the period 1980–1999. The central and right panels show the atmosphere ocean general circulation model (AOGCM) multi-model average projections for the B1 (top), A1B (middle) and A2 (bottom) emission scenarios (SRES) averaged over the decades 2020–2029 (centre) and 2090–2099 (right). The left panels show corresponding uncertainties as the relative probabilities of estimated global average warming from several different AOGCM and earth system models of intermediate complexity studies for the same periods. [Taken from IPCC (2007), SPM.6.]

Chapter 1 General introduction

(Fig. 1.1) shows scenario-independant patterns: while least warming is expected over the southern oceans and parts of the North Atlantic Ocean, temperatures are projected to increase over all continents but in particular at high northern latitudes. However, the magnitude of the temperature increase varies with the different scenarios. While for the first few decades of the 21st century the probabilistic projections of global average surface temperature change were insensitive to differences between the SRES-scenarios, much larger differences emerged by the end of the 21st century. The resulting range of global mean temperature change from AD 1990 to 2100 given by the full set of SRES scenarios is 1.4 °C to 5.8 °C (IPCC, 2007).

Climate change is not only expressed by a warming trend but also through changes in precipitation, cloudiness and the frequency of extreme events. These changes occur regionally and seasonally, affect the short-term climate variability (Edwards et al., 2007) and have serious consequences: sea ice and mountain glaciers, for example, have declined on average in both hemispheres, resulting in a considerable sea level rise. Over many large regions, long-term trends in precipitation have revealed significantly increasing precipitation in the eastern parts of North and South America as well as northern Europe and central Asia, whereas drying has been observed in the Sahel, the Mediterranean, southern Africa and parts of southern Asia. Heat waves such as the one experienced in summer 2003 in most of central Europe, where June, July and August temperatures were 5.1 °C warmer than average (Schär et al., 2004), are expected to occur at a higher frequency along with other extreme events, as for example droughts and heavy rainfall events. Although the precise magnitude and spatial pattern of future climate change is difficult to quantify, predictions based on this large number of indicators are a requisite to plan for, adapt to and mitigate the impacts of climate change in the future.

1.1 Status of paleoclimatic research

In the context of current global warming and its uncertain magnitude, reliable climate reconstructions are of particular interest. Knowledge about past climate variability assesses the amplitude of climate under natural forcings and therefore helps to place the current warming trend into a historical perspective. Paleoclimatic studies make

---
among others, as a basis for climate projections

1.1 Status of paleoclimatic research

use of changes in climatically-sensitive indicators to infer past changes in climate on different time scales ranging from decades to millions of years. Such so-called proxy-data (e.g., tree-ring width) may be influenced by local temperature but also by other factors such as precipitation and are often more representative of a particular season than the entire year (IPCC, 2007).

Proxy data can be of manifold origin including both documentary and natural archives. On a global scale, the richest sources of paleoclimate information are sediment cores from deep oceans and ice cores from the poles, but they record conditions in areas where no one lives and they lack not only in spatial, but also in temporal resolution and are therefore likely to fail in capturing rapid changes. Altough maybe not as widespread as the two latter archives, there are also many natural archives in human influenced areas, including various sediments, ice cores and trees (McCarroll and Loader, 2004; Walker and Lowe, 2007). Among these, trees provide one of the best natural archives of recent and Holocene environmental conditions. They are widespread, sensitive to their environment and produce a continuous, annually resolved biological archive. Climatically-sensitive tree-ring chronologies have already been used for a long time to reconstruct interannual climate variability on a regional scale and in combination with network approaches also spatial climate patterns up to near-hemispheric scales, extending observed climate records by several centuries (e.g., Fritts, 1976; Schweingruber, 1996; Briffa et al., 2002; Cook et al., 2004). Meanwhile, tree-ring research has been integrated worldwide into research on global change (Bradley, 1989; Eddy, 1992) and in the recent past, tree-ring derived records have played a prominent role in assessing the climate of the past. Most longer chronologies show a variability associated with local temperature and have been utilized in virtually all published studies to reconstruct Northern Hemisphere (NH) or global average surface temperature (Jansen et al., 2007; Jones et al., 2009).

Despite often providing complementary climate records, individual proxies may also suggest different pictures of past climate. Persisting controversy surrounding the pioneering NH temperature reconstruction by Mann et al. (1999)[2] underlines the importance of such records for understanding our changing climate (Esper et al., 2005b). Figure 1.2 illustrates various instrumental and proxy climate evidence of the variations in average large-scale surface temperatures over the last 1,300 years. The upper panel

---

[2]This so-called "Hockey-Stick-Curve" shows an almost linear temperature decrease from AD 1000 to the late 19th century, followed by a dramatic temperature increase up to present.

Chapter 1 General introduction

(Fig. 1.2a) shows two instrumental compilations and a mean of four European stations, which all represent the mean annual surface temperature and indicate the last two to three decades clearly as the warmest period (Jansen et al., 2007). The center panel (Fig. 1.2b) illustrates twelve temperature reconstructions (details see Table 1.1) derived from different proxies including the famous "hockey-stick" reconstruction of Mann et al. (1999), which has been the subject of several critical studies such as by Esper et al. (2002), whose reconstruction is also included. All reconstructions are expressed as temperature anomalies with respect to the 1961-1990 mean. While most archives are compiled from multiproxy evidence, including documentary evidence, physical tree-ring properties, ice cores, lake sediments and borehole temperatures (Jones et al., 1998; Mann et al., 1999; Briffa et al., 2001; Mann and Jones, 2003; Rutherford et al., 2005; Moberg et al., 2005; Hegerl et al., 2006), three are based on tree-ring evidence (Briffa, 2000; Esper et al., 2002; D'Arrigo et al., 2006), one is built from borehole temperatures (Pollack and Smerdon, 2004) and one represents the interpretation of glacier length changes (Oerlemans, 2005).

Although at first glance a wide range of archives seems to be represented, all but two studies include evidence obtained from tree rings. Tree rings, however, have greatest sensitivity to summer rather than winter conditions. Temperatures derived from summer-sensitive proxies may therefore limit the conclusions which can be drawn regarding annual temperatures (Jansen et al., 2007).

A schematic representation of the most likely hemispheric mean temperature of all reconstructions identified in Table 1.1 (except for RMO..2005 and PS2004) is highlighted in Figure 1.2c, taking into account the associated statistical uncertainty of every single record whereupon the uncertainties generally increase with time into the past due to increasingly limited spatial coverage (Jansen et al., 2007). Despite some substantial divergence during certain periods, there is coherence in relatively cool conditions in the 17th and early 19th centuries, warm conditions in the 11th and 15th centuries and warmest conditions in the 20th century. Concerning common periods, the same reconstructions show different absolute temperatures including large amplitudes (e.g., Esper et al., 2002; Pollack and Smerdon, 2004) and smaller ones (e.g., Jones et al., 1998; Mann et al., 1999), ranging from $0.4\,°C$ to $1.0\,°C$ for decadal means (Esper et al., 2005b; Moberg et al., 2005). Further, depending on the reconstruction method and instrumental target chosen, the amplitude within the same reconsuction can easily vary by more than $0.5\,°C$ (Esper et al., 2005a; Frank et al., 2007). However, although the

1.1 Status of paleoclimatic research

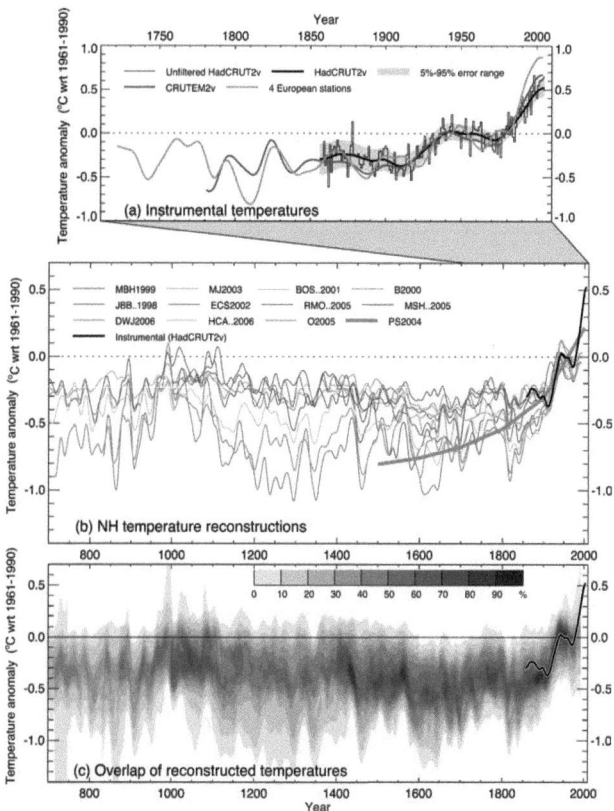

Figure 1.2: Records of Northern Hemisphere temperature variation during the last 1,300 years. All temperatures represent anomalies (°C) from the 1961 to 1990 mean. (a) Annual mean instrumental temperature records, identified in Table 1.1. (b) Reconstructions using multiple climate proxy records (Table 1.1), and the HadCRUT2v instrumental temperature record in black. (c) Overlap of the published multi-decadal time scale uncertainty ranges of all temperature reconstructions (Table 1.1) (except for RMO..2005 and PS2004), with temperatures within ± 1 standard error (SE). The HadCRUT2v instrumental temperature record is shown in black. All series have been smoothed with a Gaussian-weighted filter to remove fluctuations on time scales less than 30 years. [Taken from IPCC (2007), p. 467]

Chapter 1 General introduction

Table 1.1: Records of Northern Hemisphere temperature shown in Figure 1.2 [After IPCC (2007) p. 469]

Instrumental temperatures

Series	Period	Description	Reference
HadCRUTSv[a]	1858–2005	Land and marine temperatures for the NH	Jones and Moberg 2003; errors from Jones et al. 1997
CRUTEM2v[b]	1781–2004	Land-only temperatures for the NH	Jones and Moberg 2003; extended using data from Jones et al. 2003
4 Europ. stations	1721–2004	Average of central England, De Bilt, Berlin, Uppsala	Jones et al. 2003

Proxy-based reconstructions of temperature

Series	Period	Season	Region	Reference
JBB..1998	1000–1991	Summer	Land, NH	Jones et al. 1998; calibrated by Jones et al. 2001
MBH1999	1000–1980	Annual	Land + marine, NH	Mann et al. 1999
BOS..2001	1402–1980	Summer	Land, NH	Briffa et al. 2001
ECS2002	831–1992	Annual	Land, NH	Esper et al. 2002; recalibrated by Cook et al. 2004
B2000	1–1993	Summer	Land, NH	Briffa 2000; calibrated by Briffa et al. 2004
MJ2003	200–1980	Annual	Land + marine, NH	Mann and Jones 2003
RMO2005	1400–1960	Annual	Land + marine, NH	Rutherford et al. 2005
MSH..2005	1–1979	Annual	Land + marine, NH	Moberg et al. 2005
DWJ2006	713–1995	Annual	Land, NH	D'Arrigo et al. 2006
HCA..2006	558–1960	Annual	Land, NH	Hegerl et al. 2006
PS2004	1500–2000	Annual	Land, NH	Pollack and Smerdon 2004; reference level adjusted following Moberg et al. 2005
O2005	1600–1990	Summer	Global land	Oerlemans 2005

[a]Hadley Center/Climate Research Unit gridded surface temperature data set, version 2 variance adjusted.

[b]Climate Research Unit gridded land surface air temperature, version 2 variance corrected.

warmest conditions are identified for the 20th century by almost all reconstructions, the overall temperature discrepancies in amplitude are in the order of the total variability within the last millennium, and are therefore very likely to confuse rather than to support models of future climate (Esper et al., 2005a; Jansen et al., 2007).

In addition, there is some controversy about the so-called Medieval Warm Period (MWP) (Hughes and Diaz, 1994). Unlike Esper et al. (2005b), who identified a reasonably coherent picture of major climate episodes such as MWP, "Little Ice Age" (LIA) (Overpeck et al., 1997) and "Recent Warming" between five of these records (Jones et al., 1998; Mann et al., 1999; Briffa, 2000; Esper et al., 2002; Moberg et al., 2005), Jansen et al. (2007) revealed substantial discrepancies during the MWP concluding that the "wide spread of values exhibited by the individual record" is indicative of "the heterogeneous nature of climate during MWP". This controversy suggests that more evently distributed high-resolution proxy data are needed to assess the spatial extent of warmth during MWP Esper and Frank (2009).

1.2 Has palaeoclimate research helped? Where are we now?

Global climate models together with constraints from observations enable an assessment of climate change. To predict future changes, a simple metric such as climate sensitivity is required. Climate sensitivity is defined as the change in average global temperature after atmospheric CO_2 concentration is doubled and equilibrium is reached (Schlesinger and Mitchell, 1987) and is therefore a measure of the climate system response to sustained radiative forcing (IPCC, 2007).

The first estimate of climate sensitivity was made over a century ago by Arrhenius (1896), who calculated atmospheric temperature change with a doubling of CO_2 concentrations at various latitudes. However, more than 60 years passed before the issue of estimating climate sensitivity was revisited with the development of the first atmospheric general circulation models (GCMs) in the 1960s and 1970s (e.g., Manabe and Wetherald, 1967, 1975). Predictions of climate sensitivity made in the 1970s ranged from 1.5 °C to 4.5 °C and were summarized in a report by the National Academy of Sciences (NAS; Charney, 1979). Twenty years later, the IPCC Third Assessment Re-

Chapter 1 General introduction

port (TAR) published values in the range of 2 °C to 5.1 °C (Houghton et al., 2001). The latest IPCC report stated "It is likely[3] to be in the range 2 °C to 4.5 °C with a best estimate of about 3 °C, and is very unlikely[4] to be less than 1.5 °C. Values substantially higher than 4.5 °C cannot be excluded, but agreement of models with observations is not as good for those values." (IPCC, 2007: SPM, p.12)

Despite improvements in methodology and progress in quantifying uncertainties, the range of estimates for climate sensitivity has therefore changed surprisingly little since the first estimates were made in the 1970s. The persistence of these large uncertainties in climate sensitivity reinforces the need for further studies, because a plan for, an adaptation to, and a mitigation of possible future climate change can only be successful if the magnitude of change is known. Attempts in constraining climate sensitivity using paleoclimatic data (e.g., by constraining climate models, see Chapter 1.3) have not yet been satisfactory, as the uncertainty range of estimates was not substantially reduced. This may be mainly due to the fact that the uncertainties associated with the forcings themselves are larger than those associated with recent changes in forcing, but also suggest a less-than-optimal use of paleoclimate data. Nevertheless, recent paleoclimate work indicates that there is potential to improve the robustness of constraints on climate sensitivity derived from paleoclimate data, by (1) increasing the number of proxies in spatial coverage and temporal resolution and (2) by improving the understanding of the nature of processes, both climatic and non-climatic, that influence climate proxy data so that limitations within the proxy series can be recognized (Edwards et al., 2007; Walker and Lowe, 2007; Jones et al., 2009).

Regardless of the variety of proxy data, most climate reconstructions are focused on temperature. Climate change, however, is not only expressed in rising temperature but also in changes in the regional to continental circulation patterns and the frequency and intensity of droughts and floods (Jones et al., 2009). As precipitation plays a key role for human livelihood and economies as well as for many terrestrial ecosystems (Pauling et al., 2006), knowledge concerning the past range of variability in hydroclimatic variables such as regional precipitation and the occurence of extreme events provides a valuable complement to temperature reconstructions and is therefore of particular interest. However, hydroclimatic reconstructions are extremely rare in central and northern Europe, among others because of the local nature of moisture sig-

---
[3]Likelihood of occurrence 66–90%
[4]Likelihood of occurrence 1–10%

nals. Consequently, the interest in such records may help to develop and optimize new proxies, such as stable isotope series from various materials, which have been largely underrepresented so far (Leng, 2005).

1.3 The Millennium project

The EU funded Millennium project (2006-2010, http://geography.swansea.ac.uk/millennium/) is a multidisciplinary consortium of more than 39 European universities and research institutes, with the aim of answering one single question:

> "Does the magnitude and rate of 20th century climate change exceed the natural variability of European climate over the last millennium?"

To address this question soundly, an accurate picture of European climate over the last one thousand years is needed. Currently such a picture does not exist. Despite many attempts to reconstruct the past climate of the NH and Europe (see Fig 1.2 and Tab 1.1), results vary in particular with respect to longer-term changes. Figure 1.3 stresses some of the current limitations in understanding past European climate and shows how Millennium will address them. The present state of climate research reveals some reasons why large-scale climate reconstructions fail to portray the true amplitude of past temperature change:

i the lack of high resolution millennial-long climate reconstructions (e.g., Moberg et al., 2005; Esper and Frank, 2009);

ii the bias of regression-based calibration techniques that under-estimate amplitude change, and a general uncertainty in differing calibration approaches applied in the recent literature (e.g., von Storch et al., 2004; Esper et al., 2005a);

iii the limitation of tree-ring based climate reconstructions to retain the full spectrum of lower frequency climate variation (e.g., Cook et al., 1995);

iv the limited coverage of long instrumental records and the systematic bias noted in some of the early instrumental data used to calibrate proxy-based reconstructions (e.g., Böhm et al., 2001; Frank et al., 2007).

Chapter 1 General introduction

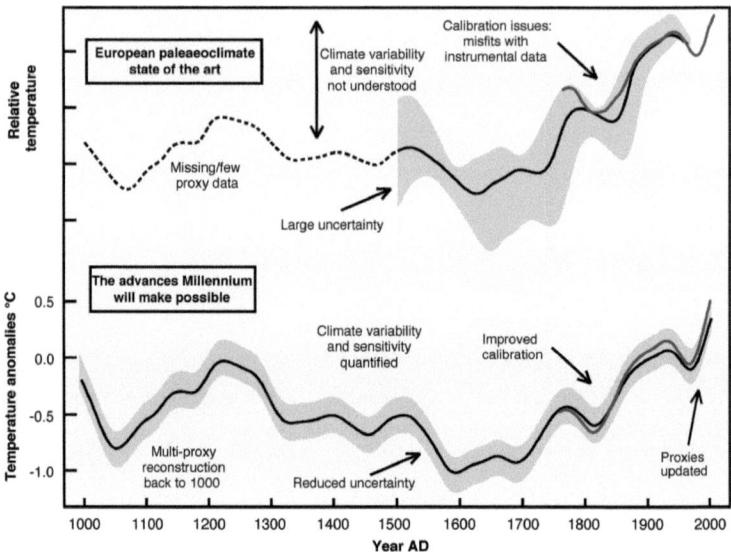

Figure 1.3: Schematic diagramm showing the current limitations of European palaeoclimate reconstructions and the advancements in the detection and quantitative understanding of European climate variability that Millennium will make possible [provided by Esper, J.; Wilson, R.S.J.; Büntgen, U. (pers. comm., 2005)]

Overcoming all these substantial limitations within one project will be difficult. However, as long as the absolute amplitude of past temperature variability is uncertain (see Fig. 1.1), the predictive skill of climate models is limited. Therefore, Millennium is focussing its investigations on Europe where previous research has shown a more varied climate compared to large-scale NH reconstructions and is thereby concentrating on the last 1,000 years, stressing six particular aims:

- building a database of the most reliable data on past climate;

- producing new millennial-length palaeoclimate data using innovative methods;

- combining existing and new data to reconstruct the climate of Europe for the last one thousand years at a range of spatial scales;

- using the reconstructions to define the natural variability of European climate, over both space and time, and taking account of changes in seasonality;

- testing the ability of climate models to reproduce the magnitude of natural climate variability in the past;

- predicting the probability of European climate passing critical thresholds, taking full account of the natural variability as well as greenhouse gas forcing.

By realizing these aims, in particular by generating new millennial-length reconstructions using a range of instrumental, documentary and natural archives, Millennium is likely to fully address limitations (i) and (iii). In addition, if Millennium compiles a variety of proxies with different errors with a high replication to reduce the signal-to-noise ratio and applies strong- and unbiased calibrations with quantified uncertainty, there will be substantial contributions to minimize limitations (ii) and (iv). Furthermore, these new data-sets will contribute to improve model predictions of future climate change and its impacts.

The work of Millennium is divided between five subgroups, four of which focus on different direct and proxy sources of information on past climate. The fifth group deals with data analysis and modeling. **Subgroup 1** is dealing with instrumental and documentary archives, to obtain an index of seasonal temperature and precipitation. Potential sources include chronicles, ship logs, port records, and a wide range of administrative documents, such as for example grape harvest dates, from all over Europe. **Subgroup 2** is using tree-ring evidence, which includes traditional archives such as tree-ring width and maximum latewood density but also stable isotopes of carbon, oxygen and hydrogen. **Subgroup 3** is analyzing lake sediments, peat mires and ice cores in great detail, to obtain information on subtle changes in chemistry and the microscopic remains of animals and plants. **Subgroup 4** is dealing with marine sediments and long-lived clams to provide a high-resolution record of past changes in the marine environment. **Subgroup 5** is in charge of compiling new and existing proxy records at a variety of temporal resolutions and spatial scales to test and improve models that will be ultimately used to make more refined estimates of future climate change.

Chapter 1 General introduction

To date, many new records have been compiled and many exciting findings have been made. Furthermore, new methods have been developed and tested in order to facilitate access to special archives or to reduce the amount of effort required to make reconstructions. One prominent example is the minimum blue intensity measurements of tree rings using a flat-bed scanner and commercially available software to obtain a robust and reliable surrogate for maximum latewood density, while being inexpensive and accessible (Campbell et al., 2007). Regarding the numerous interesting new records, two representative examples shall be mentioned. The first example is the winter/spring temperature reconstruction for Stockholm, which is based on harbor tax records (Leijonhufvud et al., 2008, currently extending back to AD 1500). This record is of particular interest as it represents an absolute temperature measure (freezing of Stockholm harbor) on the one hand and winter/spring temperature on the other hand, while most of the other proxies are restricted to summer season and a measure of relative temperatures. The second example are two chronologies from the clam *Arctica islandica* also called "the tree of the seas", which provide an annually banded shell. The shell's growth increments were cross-dated successfully and the discovery of some very long-lived individuals enabled the extension of the North Icelandic Shelf chronology to nearly 700 years (Stott et al., 2010), and the chronology of the Irish Sea has so far resulted in a 500-year chronology (Butler et al., 2010). The band widths of these shells carry a strong common signal that can be calibrated to temperature. This annually-resolved archive is unique in the marine environment and will contribute to a more reliable dating of other chronologies from marine sediments.

The relevance of these reconstructions in constraining models predicting future climate change is demonstrated in Figure 1.4, in which climate predictions were made by numerous runs of simple energy balance models with different internal parameters. While in Figure 1.4a the climate projections are based on instrumental observations only, each model run in Figure 1.4b used a proxy-based temperature reconstruction and recent instrumental observations to assign similarity scores for both pre- and post-instrumental periods. In addition, uncertainties in the models' internal parameters, external forcings, temperature reconstruction, and recent instrumental observations were taken into account. The overall uncertainty was then calculated by combining the similarity score and the uncertainty scores for each simulation and projected into the future. The results of this modeling exercise suggest that the proxy-based climate reconstruction imposes a strong additional constraint on the conventional climate projection that is solely based on instrumental observations (Yamazaki et al., 2009).

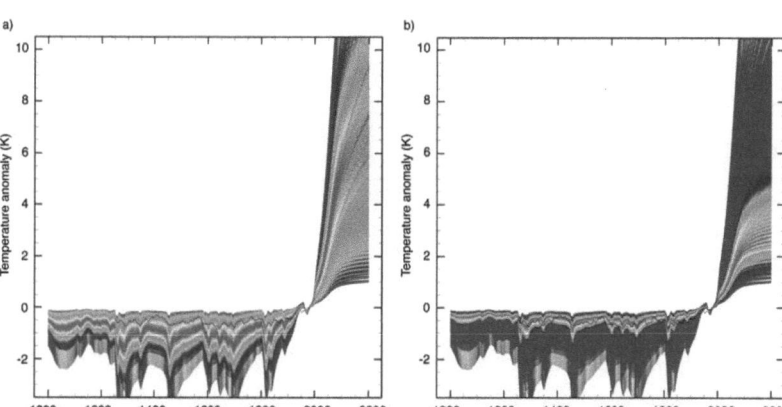

Figure 1.4: Relative likelihood of temperature expressed as anomalies in Kelvin (K) without (a) and with palaeoclimatic constraints (b). The color code ranges from very likely (red) to very unlikely (purple) [provided by Yamazaki et al., 2009].

Finally, to address future climate change at a high temporal and spatial resolution, Millennium will use a general circulation model (GCM), incorporating the geography of climate change, and better uncertainty estimates for the palaeoclimate reconstructions (see also http://www.climateprediction.net). The project is ongoing and it remains exciting to see which picture of Europe's past climate will be revealed.

1.4 This study: objectives and research questions

This PhD thesis represents a part of the Millennium project. Accordingly, its overall goal was to build a stable carbon and oxygen isotope chronology of more than 1,000 years from tree-ring cellulose of European larch (*Larix decidua* Mill.) from Alpine sites and to explore the driving factors of stable isotope signatures in these chronologies.

Chapter 1 General introduction

European larch is a species of trees growing at the highest altitudes in the European Alps and also one of the most temperature sensitive in this zone. With a longevity of 850+ years, its Alps-wide utilization as timber and its high temperature sensitivity, it is considered to be an ideal archive for climate reconstructions. Thus, the isotopic signatures of its tree-ring cellulose are likely to reflect climate variability to a certain extent.

Establishing these long isotope chronologies involved the sampling of both living trees and historical timber from the Lötschental and Simplon region (Valais, Switzerland), dating each ring to its year of formation, separating every single tree ring, extraction of tree-ring alpha-cellulose and stable isotope analysis. This work was accomplished in collaboration with the dendro-unit of the Swiss Federal Research Institute for Forest, Snow and Landscape Research (WSL) and the Division of Climate and Environmental Physics of the University of Bern.

As the Alpine region is highly sensitive to climatic changes, understanding its natural climate variability is an important component of a comprehensive picture of Europe's past climate and a prerequisite to correctly interpret the trends of the last decades under growing anthropogenic influence. In this context, these stable-isotope chronologies can help to constrain climate models and should therefore contribute to improved projections of future climate change.

From an ecological point of view, the focus was set on the following questions:

- Are reliable long-term climate reconstructions from tree-ring stable isotopes of European larch (*Larix decidua* Mill.) feasible?
- What is the dominating climate signal in stable carbon and oxygen isotope series from Alpine sites?
- Is there a biological bias that can disturb the climate signal in the isotopic signatures and may therefore cause errors in climate reconstructions?

To answer these questions, the following challenges had to be addressed:

- Is the isotopic signature in earlywood different from the one in latewood?
- Does the occurrence of Grey larch budmoth (*Zeiraphera diniana* Gn.) influence the isotopic signature in tree rings?

- Do isotopic signatures in living wood differ from those in historical material? Is it therefore feasible to link cohorts of different origin?

1.5 Thesis structure

Chapter 2: Stable isotopes in dendroclimatology

This chapter represents an extended introduction and gives a more detailed overview of the particular application of stable carbon and oxygen isotopes as palaeoclimatic archives. It includes a brief history how stable isotopes became a palaeoclimatic proxy as well as the present knowledge in carbon and oxygen isotope theory and their variation in plants. In addition, the state-of-the-art in climate reconstructions based on stable carbon and oxygen isotopes is presented, so that the following chapters can be interpreted and discussed in this context.

Chapter 3: Stable isotope coherence in the earlywood and latewood of tree-line conifers

Annually resolved and replicated tree-ring stable isotope series have the potential to allow for a reconstruction of climate parameters, such as temperature, at multi-millennial timescales and at all temporal frequencies (e.g., McCarroll and Loader, 2004). Over such timescales, sample numbers are large and preparation is costly and time consuming. Furthermore, the impact of storage and remobilization of photosynthates on the isotopic composition of the resulting tree ring is not completely understood and is a subject of ongoing research (e.g., Helle and Schleser, 2004). To date, many studies have utilised only latewood rather than the entire tree ring to maximise the annual resolution and to avoid carry-over effects from reserves (e.g., Gagen et al., 2007; Etien et al., 2008). However, the old trees from tree-line locations that are required to build long chronologies often display narrow ring-widths (< 0.5 mm), making accurate earlywood-latewood separation difficult, often inaccurate and time consuming. In this study, tree-ring stable isotopes from two European tree-line locations were analysed to test the homogeneity of signals between latewood, earlywood and whole-ring cellulose. The results from Scots pine (*Pinus sylvestris* L.) from northern Norway and European larch (*Larix decidua* Mill.) from southern Switzerland demonstrate a

Chapter 1 General introduction

high common signal between isotopes from earlywood and latewood in both species and suggest that, for European tree-line conifers, the separation of earlywood from latewood is not necessary to resolve an annual isotopic signal. The following analyses are therefore all based on whole ring cellulose.

Chapter 4: Summer temperature dependency of larch budmoth outbreaks revealed by an Alpine stable isotope chronology

The Grey larch budmoth (*Zeiraphera diniana* Gn.) is a foliage-feeding insect that is characterized by periodical outbreaks, causing discernable physical alterations of cell growth in tree rings of European larch (*Larix decidua* Mill.) in the European Alps (Baltensweiler et al., 1977). However, it remains unclear how these population cycles are modulated by climatic influences (Esper et al., 2007) and if they also impact isotopic signatures in tree-ring cellulose, thereby masking climate signals. To prevent errors in potential climate reconstructions, this needs to be determined. Thus, the possible influence of larch budmoth on the isotopic signature of tree rings was investigated by comparing outbreak events in stable carbon and oxygen isotope chronologies of larch and their corresponding tree-ring widths from two high-elevation sites (1800–2200 m a.s.l.) in the Swiss Alps against isotope data obtained from non-host spruce (*Picea abies*). In addition, a comparison was made with long-term monthly resolved temperature data. This comparison surprisingly revealed a strong coherence between summer temperatures and larch budmoth defoliation events, but provided no evidence of disturbed isotope signatures during these events. Within the constraints of this rather local study, we conclude that isotopic ratios in tree rings of larch provide a strong and consistent temperature signal, which is unbiased by larch budmoth outbreaks. Stable isotopes in tree-ring cellulose of larch therefore not only enable the analysis of climate-driven changes of budmoth cycles in the long term, but they also are an excellent climate proxy unbiased by insect disturbances. Thus, the isotope series used in the following application studies do not need any corrections for these outbreak events.

Chapter 5: A 350-year drought reconstruction from Alpine tree-ring stable isotopes

The potential of tree-ring stable isotopes for quantitative terrestrial palaeoclimatic reconstruction has been demonstrated in numerous studies (Robertson et al.,

1.5 Thesis structure

2008, and references therein). However, climate records obtained from stable isotopes are still rather controversial (see Chapter 2.2). This may be due to the fact that climate signals in isotope series can be highly site dependent and a potentially multi-parameter climate signal cannot be excluded. As shown in Chapter 3 and 4, the Lötschental stable isotope series from European larch (*Larix decidua* Mill.) indicate a strong climate forcing unbiased by biological disturbances and systematic errors caused by earlywood/latewood separation. Hence, in a detailed sensitivity study, different climate-isotope relationships were investigated considering varying temporal and spatial resolutions. Unlike tree-ring width and maximum latewood density, which contain only summer temperature information at this site (Büntgen et al., 2005, 2006), the carbon isotopes reveal – in addition to a strong temperature signal – a striking sensitivity to precipitation while in the oxygen isotopes sunshine duration is clearly the dominating climate factor. Based on these analyses, we present the first carbon-isotope derived drought reconstruction for the European Alps, which provides new evidence for inter-annual to long-term changes in central European summer moisture variability from AD 1650–2004. The climate-isotope relationships revealed in this study provide the basis for the interpretation of the millennial isotope chronologies in the following study.

Chapter 6: 1200 years of climate history from stable carbon and oxygen isotopes

Twentieth-century warming is accompanied by altering temporal and spatial distribution of precipitation (Jansen et al., 2007). Such changes may affect human well-being and ecosystem dynamics even more strongly than the rising temperature itself (Kundzewicz et al., 2007). While the past, present and projected rates of regional to global temperature have been investigated extensively, only little is known about past modifications of the hydrological cycle (Huntington, 2006). Here we present the longest available annually resolved tree-ring records of carbon and oxygen isotopes in central Europe. These records from European larch (*Larix decidua* Mill.) in the Swiss Alps provide millennial-scale information on mid- to late-summer moisture variability (carbon) and sunshine duration (oxygen). According to the sensitivity study performed in Chapter 5, these two series are highly climate-sensitive with regional extent and are therefore likely to complement central Europe's picture of past climate variability.

Chapter 1 General introduction

Along with a maximum latewood density-based temperature reconstruction from the same site (Büntgen et al., 2006), we are able to subdivide climate periods such as the "Little Ice Age" and the "Medieval Warm Period" in wet to dry and sunny to cloudy episodes. This knowledge, in particular about past moisture variability, should help to constrain climate predictions and will therefore contribute to plan for, adapt to and mitigate impacts of future climate change.

Chapter 7: Conclusions and outlook

In this chapter, the results of Chapter 3 to 6 are discussed in the context of the present knowledge in stable isotope dendroclimatology (Chapter 2). The emphasis of this synthesis chapter is on the advances this study is contributing to stable isotope dendroclimatology, but it also includes a discussion of the limitations of this work and provides perspectives of future research.

References

Arrhenius, S. (1896). On the influence of carbonic acid in the air upon the temperature of the ground. *Philosophical Magazine and Journal of Science*, 41:239–276.

Baltensweiler, W., Benz, G., Bovey, P., and Delucchi, V. (1977). Dynamics of larch bud moth populations. *Annual Review of Entomology*, 22:79–100; doi:10.1146/annurev.en.22.010177.

Böhm, R., Auer, I., Brunetti, M., Maugeri, M., Nanni, T., and Schoner, W. (2001). Regional temperature variability in the European Alps: 1760-1998 from homogenized instrumental time series. *International Journal of Climatology*, 21(14):1779–1801; doi:10.1002/joc.689.

Bonan, G. (2002). *Ecological climatology*. Cambridge University Press, New York.

Bradley, R., editor (1989). *Global canges in the past. Papers arising from the 1989 Symp. Global Change Inst.* UCAR/Office for Interdisciplinery Earth Studies, Boulder, Colorado.

Briffa, K. R. (2000). Annual climate variability in the Holocene: interpreting the message of ancient trees. *Quaternary Science Reviews*, 19:87–105; doi:10.1016/S0277-3791(99)00056-6.

Briffa, K. R., Osborn, T. J., and Schweingruber, F. H. (2004). Large-scale temperature inferences from tree rings: a review. *Global Planetary Change*, 40(1-2):11–26; doi:10.1016/S0921-8181(03)00095-X.

Briffa, K. R., Osborn, T. J., Schweingruber, F. H., Harris, I. C., Jones, P. D., Shiyatov, S. G., and Vaganov, E. A. (2001). Low-frequency temperature variations from a northern tree ring density network. *Journal of Geophysical Research*, 106(D3):2929–2941; doi:10.1029/2000JD900617.

Briffa, K. R., Osborn, T. J., Schweingruber, F. H., Jones, P. D., Shiyatov, S. G., and Vaganov, E. (2002). Tree-ring width and density data around the Northern Hemisphere: Part 1, local and regional climate signals. *The Holocene*, 12(6):737–757; doi:10.1191/0959683602hl587rp.

Büntgen, U., Esper, J., Frank, D. C., Nicolussi, K., and Schmidhalter, M. (2005). A 1052-year tree-ring proxy for Alpine summer temperatures. *Climate Dynamics*, 25(2-3):141–153; doi:10.1007/s00382-005-0028-1.

Chapter 1 General introduction

Büntgen, U., Frank, D. C., Niervergelt, D., and Esper, J. (2006). Summer temperature variations in the European Alps, AD 755-2004. *Journal of Climate*, 19(21):5606–5623; doi:10.1175/JCLI3917.1.

Butler, P. G., Richardson, C. A., Scourse, J. D., Wanamaker, A. D. J., Shammon, T. M., and Bennell, J. D. (2010). Marine climate in the Irish Sea: analysis of a 489-year marine master chronology derived from growth increments in the shell of the clam *Arctica islandica*. *Quaternary Science Reviews*, 29(13-14):1614–1632; doi:10.1016/j.quascirev.2009.07.010.

Campbell, R., McCarroll, D., Loader, N. J., Grudd, H., Robertson, I., and Jalkanen, R. (2007). Blue intensity in *Pinus sylvestris* tree-rings: developing a new palaeoclimate proxy. *The Holocene*, 17(6):821–828; doi:10.1177/0959683607080523.

Charney, J. (1979). *Carbon dioxide and climate: a scientific assessment*. National Academy of Sciences, Washington, DC.

Chylek, P. and Lohmann, U. (2008). Aerosol radiative forcing and climate sensitivity deduced from the Last Glacial Maximum to Holocene transition. *Geophysical Research Letters*, 35:L04804; doi:10.1029/2007GL032759.

Cook, E. R., Briffa, K. R., Meko, D. M., Graybill, D. A., and Funkhouser, G. (1995). The segment length curse in long tree-ring chronology development for paleoclimatic studies. *The Holocene*, 5(2):229–237; doi:10.1177/095968369500500211.

Cook, E. R., Esper, J., and D'Arrigo, R. D. (2004). Extra-tropical Northern Hemisphere land temperature variability over the past 1000 years. *Quaternary Science Reviews*, 23(20-22):2063–2074; doi:10.1016/j.quascirev.2004.08.013.

D'Arrigo, R., Wilson, R., and Jacoby, G. (2006). On the long-term context for late twentieth century warming. *Journal of Geophysical Research*, 111(D3):D03103; doi:10.1029/2005JD006352.

Eddy, A., editor (1992). *The PAGES project proposed implementation plans for research activities*, Stockholm. Global Change Report 19.

Edwards, T. L., Crucifix, M., and Harrison, S. P. (2007). Using the past to constrain the future: how the palaeorecord can improve estimates of global warming. *Progress in Physical Geography*, 31:481–500; doi:10.1177/0309133307083295.

Esper, J., Büntgen, U., Frank, D. C., Niervergelt, D., and Liebhold, A. (2007). 1200 years of regular outbreaks in alpine insects. *Proceedings of the Royal Society B*, 274:671–679; doi:10.1098/rspb.2006.0191.

Esper, J., Cook, E. R., and Schweingruber, F. H. (2002). Low-frequency signals in long tree-ring chronologies for reconstructing past temperature variability. *Science*, 295(5563):2250–2253; doi10.1126/science.1066208.

Esper, J. and Frank, D. (2009). The IPCC on a heterogeneous Medieval Warm Period. *Climatic Change*, 94(3):267–273; doi:10.1007/s10584-008-9492-z.

Esper, J., Frank, D. C., Wilson, R. J. S., and Briffa, K. R. (2005a). Effect of scaling and regression on reconstructed temperature amplitude for the past millennium. *Geophysical Research Letters*, 32(7):L07711; doi:10.1029/2004GL021236.

Esper, J., Wilson, R. J. S., Frank, D. C., Moberg, A., Wanner, H., and Luterbacher, J. (2005b). Climate: past ranges and future changes. *Quaternary Science Reviews*, 24(20-21):2164–2166; doi:10.1016/j.quascirev.2005.07.001.

Etien, N., Daux, V., Masson-Delmotte, V., Stievenard, M., Bernard, V., Durost, S., Guillemin, M. T., Mestre, O., and Pierre, M. (2008). A bi-proxy reconstruction of Fontainebleau (France) growing season temperature from AD 1596 to 2000. *Climate of the Past*, 4(2):91–106.

Frank, D., Buntgen, U., Bohm, R., Maugeri, M., and Esper, J. (2007). Warmer early instrumental measurements versus colder reconstructed temperatures: shooting at a moving target. *Quaternary Science Reviews*, 26(25-28):3298–3310; doi:10.1016/j.quascirev.2007.08.002.

Fritts, H. C. (1976). *Tree Rings and Climate*. Academic Press, London, England.

Gagen, M., McCarroll, D., Loader, N. J., Robertson, L., Jalkanen, R., and Anchukaitis, K. J. (2007). Exorcising the 'segment length curse': summer temperature reconstruction since AD 1640 using non-detrended stable carbon isotope ratios from pine trees in northern Finland. *The Holocene*, 17(4):435–446; doi:10.1177/0959683607077012.

Hegerl, G. C., Crowley, T. J., Hyde, W. T., and Frame, D. J. (2006). Climate sensitivity constrained by temperature reconstructions over the past seven centuries. *Nature*, 440:1029–1032; doi:10.1038/nature04679.

Chapter 1 General introduction

Helle, G. and Schleser, G. H. (2004). Beyond CO_2-fixation by Rubisco - an interpretation of $^{13}C/^{12}C$ variations in tree rings from novel intra-seasonal studies on broad-leaf trees. *Plant, Cell and Environment*, 27(3):367–380; doi:10.1111/j.0016-8025.2003.01159.x.

Houghton, J. T., Ding, Y., Griggs, D. J., Noguer, M., van der Linden, P. J., and Xiaosu, D., editors (2001). *Climate Change 2001: the Physical Science Basis. Contribution of Working Group I to the Third Assessment Report of the Intergovernmental Panel on Climate Change*. Cambridge University Press, Cambridge, United Kingdom and New York, NY, USA.

Hughes, M. K. and Diaz, H. F. (1994). Was there a "Medieval Warm Period", and if so, where and when? *Climatic Change*, 26(2-3):109–142; doi:10.1007/BF01092410.

Huntington, T. G. (2006). Evidence for intensification of the global water cycle: review and synthesis. *Journal of Hydrology*, 319(1-4):83–95; doi:10.1016/j.jhydrol.2005.07.003.

IPCC (2007). *Climate Change 2007: the Physical Science Basis. Contribution of Working Group I to the Fourth Assessment Report of the Intergovernmental Panel on Climate Change*. Cambridge University Press, Cambridge, United Kingdom and New York, NY, USA.

Jansen, E., Overpeck, J., Briffa, K. R., Duplessy, J.-C., Joos, F., Masson-Delmotte, V., Olago, D., Otto-Bliesner, B., Peltier, W. R., Rahmstorf, S., Ramesh, R., Raynaud, D., Rind, D., Solomina, O., Villalba, R., and Zhang, D. (2007). Palaeoclimate. In Solomon, S., Qin, D., Manning, M., Chen, Z., Marquis, M., Averyt, K., Tignor, M., and Mille, H., editors, *Climate Change 2007: The Physical Science Basis. Contribution of Working Group I to the Fourth Assessment Report of the Intergovernmental Panel on Climate Change*, pages 433–497. Cambridge University Press, Cambridge, United Kingdom and New York, NY, USA.

Jones, P. D., Briffa, K. R., Barnett, T. P., and Tett, S. F. B. (1998). High-resolution palaeoclimatic records for the last millennium: interpretation, integration and comparison with general circulation model control-run temperatures. *The Holocene*, 8(4):455–471; doi:10.1191/095968398667194956.

Jones, P. D., Briffa, K. R., and Osborn, T. J. (2003). Changes in the Northern

Hemisphere annual cycle: implications for paleoclimatology? *Journal of Geophysical Research*, 108(D18):4588; doi:10.1029/2003JD003695.

Jones, P. D., Briffa, K. R., Osborn, T. J., Lough, J. M., van Ommen, T. D., Vinther, B. M., Luterbacher, J., Wahl, E. R., Zwiers, F. W., Mann, M. E., Schmidt, G. A., Ammann, C. M., Buckley, B. M., Cobb, K. M., Esper, J., Goosse, H., Graham, N., Jansen, E., Kiefer, T., Kull, C., Küttel, M., Mosley-Thompson, E., Overpeck, J. T., Riedwyl, N., Schulz, M., Tudhope, A., Villalba, R., Wanner, H., Wolff, E., and Xoplaki, E. (2009). High-resolution palaeoclimatology of the last millennium: a review of current status and future prospects. *The Holocene*, 19(1):3–49; doi:10.1177/0959683608098952.

Jones, P. D. and Moberg, A. (2003). Hemispheric and large-scale surface air temperature variations: an extensive revision and an update to 2001. *Journal of Climate*, 16(2):206–223; doi:10.1175/1520–0442(2003)016<0206:HALSSA>2.0.CO;2.

Jones, P. D., Osborn, T. J., and Briffa, K. R. (1997). Estimating sampling errors in large-scale temperature averages. *Journal of Climate*, 10(10):2548–2568; doi:10.1175/1520–0442(1997)010<2548:ESEILS>2.0.CO;2.

Jones, P. D., Osborn, T. J., and Briffa, K. R. (2001). The evolution of climate over the last millennium. *Science*, 292(5517):662–667; doi:10.1126/science.1059126.

Kundzewicz, Z., Mata, L., Arnell, N., Döll, P., Kabat, P., Jiminéz, B., Miller, K., Oki, T., Sen, Z., and Shiklomanov, I. (2007). Freshwater resources and their management. In Parry, M., Canziani, O., Palutikof, J., van der Linden, P. J., and Hanson, C., editors, *Climate Change 2007: Impacts, Adaptation and Vulnerybility. Contribution of Working Group II to the Fourth Assessment Report of the Intergovernmental Panel on Climate Change*, pages 173–210. Cambridge University Press, Cambridge, United Kingdom and New York, NY, USA.

Lamb, H. (2005). *Climate, History and the Modern World*. Routledge, London.

Leijonhufvud, L., Wilson, R., and Moberg, A. (2008). Documentary data provide evidence of Stockholm average winter to spring temperatures in the eighteenth and nineteenth centuries. *The Holocene*, 18(2):333–343; doi:10.1177/0959683607086770.

Leng, M. J., editor (2005). *Isotopes in Palaeoenvironmental Research*, volume 10 of *Developments in Palaeoenvironmental Research Series*. Springer, Dordrecht.

Manabe, S. and Wetherald, R. T. (1967). Thermal equilibrium of atmosphere with a given distribution of relative humidity. *Journal of the Atmospheric Sciences*, 24(3):241–259; doi:10.1175/1520-0469(1967)024.

Manabe, S. and Wetherald, R. T. (1975). Effects of doubling CO_2 concentrations on climate of a general circulation model. *Journal of the Atmospheric Sciences*, 32(1):3–15; doi:10.1175/1520-0469(1975)032<0003:TEODTC>2.0.CO;2.

Mann, M. E., Bradley, R. S., and Hughes, M. K. (1999). Northern Hemisphere temperatures during the past millennium: inferences, uncertainties, and limitations. *Geophysical Research Letters*, 26(6):759–762; doi:10.1029/1999GL900070.

Mann, M. E. and Jones, P. D. (2003). Global surface temperatures over the past two millennia. *Geophysical Research Letters*, 30(15):1820; doi:10.1029/2003GL017814.

McCarroll, D. and Loader, N. J. (2004). Stable isotopes in tree rings. *Quaternary Science Reviews*, 23(7-8):771–801; doi:10.1016/j.quascirev.2003.06.017.

Moberg, A., Sonechkin, D. M., Holmgren, K., Datsenko, N. M., and Karlén, W. (2005). Highly variable Northern Hemisphere temperatures reconstructed from low- and high-resolution proxy data. *Nature*, 433(7026):613–617; doi:10.1038/nature03265.

Oerlemans, J. (2005). Extracting a climate signal from 169 glacier records. *Science*, 308(5722):675–677; doi:10.1126/science.1107046.

Overpeck, J., Hughen, K., Hardy, D., Bradley, R., Case, R., Douglas, M., Finney, B., Gajewski, K., Jacoby, G., Jennings, A., Lamoureux, S., Lasca, A., MacDonald, G., Moore, J., Retelle, M., Smith, S., Wolfe, A., and Zielinski, G. (1997). Arctic environmental change of the last four centuries. *Science*, 278(5341):1251–1256; doi:10.1126/science.278.5341.1251.

Pauling, A., Luterbacher, J., Casty, C., and Wanner, H. (2006). Five hundred years of gridded high-resolution precipitation reconstructions over Europe and the connection to large-scale circulation. *Climate Dynamics*, 26(4):387–405; doi:10.1007/s00382-005-0090-8.

Petit, J. R., Jouzel, J., Raynaud, D., Barkov, N. I., Barnola, J. M., Basile, I., Bender, M., Chappellaz, J., Davis, M., Delaygue, G., Delmotte, M., Kotlyakov, V. M., Legrand, M., Lipenkov, V. Y., Lorius, C., Pepin, L., Ritz, C., Saltzman, E., and Stievenard, M. (1999). Climate and atmospheric history of the past 420,000 years from the Vostok ice core, Antarctica. *Nature*, 399(6735):429–436; doi:10.1038/20859.

Pollack, H. N. and Smerdon, J. E. (2004). Borehole climate reconstructions: spatial structure and hemispheric averages. *Journal of Geophysical Research*, 109(D11):D11106; doi:10.1029/2003JD004163.

Robertson, I., Leavitt, S. W., Loader, N. J., and Buhay, W. (2008). Progress in isotope dendroclimatology. *Chemical Geology*, 252(1-2):EX1–EX4; doi:10.1016/s0009-2541(08)00177-0.

Rutherford, S., Mann, M. E., Osborn, T. J., Bradley, R. S., Briffa, K. R., Hughes, M. K., and Jones, P. D. (2005). Proxy-based Northern Hemisphere surface temperature reconstructions: sensitivity to method, predictor network, target season, and target domain. *Journal of Climate*, 18(13):2308–2329; doi:10.1175/JCLI3351.1.

Schär, C., Vidale, P. L., Luthi, D., Frei, C., Haberli, C., Liniger, M. A., and Appenzeller, C. (2004). The role of increasing temperature variability in European summer heatwaves. *Nature*, 427(6972):332–336; doi:10.1038/nature02300.

Schlesinger, M. E. and Mitchell, J. F. B. (1987). Climate model simulations of the equilibrium climatic response to increased carbon dioxide. *Reviews of Geophysics*, 25(4):760–798; doi:10.1029/RG025i004p00760.

Schweingruber, F. H. (1996). *Tree Rings and Environment - Dendroecology*. Paul Haupt, Berne, Stuttgart, Vienna.

Stott, K. J., Austin, W. E. N., Sayer, M. D. J., Weidman, C. R., and Wilson, R. J. S. (2010). The potential of *Arctica islandica* growth records to reconstruct coastal climate in north west Scotland, UK. *Quaternary Science Reviews*, 29(13-14):1602–1613; doi:10.1016/j.quascirev.2009.06.016.

von Storch, H., Zorita, E., Jones, J. M., Dimitriev, Y., Gonzalez-Rouco, F., and Tett, S. F. B. (2004). Reconstructing past climate from noisy data. *Science*, 306(5696):679–682; doi:10.1126/science.1096109.

Walker, M. and Lowe, J. (2007). Quaternary science 2007: a 50-year retrospective. *Journal of Geological Society*, 164:1073–1092; doi:10.1144/0016–76492006-195.

Yamazaki, Y. H., Allen, M. R., Huntingford, C., Frame, D. J., and Frank, D. C. (2009). Refining future climate projections using uncertain climate data of the last millennium. In *Geophysical Research Abstracts*, volume 11. EGU General Assembly 2009.

2

Stable isotopes in dendroclimatology

Dendroclimatology is an established field of paleoclimatology and has been widely used since the early 20th century. The most common way to extract a paleoclimate signal from tree rings is to measure the width of the ring, but all physical properties of tree rings (width, density and reflectance) represent a record of past environmental changes, which may be used to extract palaeoclimatic information at annual resolution (Fritts, 1976; Schweingruber, 1983, 1996). Suitable trees are widespread and by piecing together the records from living, dead and subfossil wood it is possible to produce continuous, perfectly dated chronologies throughout the Holocene (Eronen et al., 2002; Grudd et al., 2002; Hantemirov and Shyatov, 2002; Leuschner et al., 2002; Spurk et al., 2002). As physical tree-ring properties have proved to be reliable proxies for climate reconstruction at sites where tree growth is limited by one dominant climate factor (Esper et al., 2002), these long tree-ring chronologies provide the basis for much of our present knowledge about the Earth's past climate variability (Briffa, 2000; Jones and Mann, 2004; Esper et al., 2005).

Although ring-width chronologies deliver some of the best high-resolution paleoclimate datasets available, there are a number of limitations that restrict their potential. One of the most prominent constraints has become known as the so-called "segment length curse" and derives from the fact that, during tree growth, the ring widths tend to decline due to the increasing circumference of the trunk (Cook et al., 1995). This decline can be statistically removed by using regression-based techniques, but climate signals occurring at the scale of the life-span of the tree (the segment length) will be affected as well. Alternative detrending methods (e.g., regional curve standardization) attempt to retain a greater proportion of the low frequency information, but require very large data sets (Cook et al., 1995; Esper et al., 2002). Another significant problem is that strong climate-ring-width relationships are restricted to sites where tree growth is limited by one dominant climatic factor (e.g., temperature at tree-line sites). In more mesic climates such as in most of central Europe, the climatic influence on tree growth is rather complex and no simple climate-growth relationship can be identified (Saurer et al., 1995). By using stable isotopes it may be possible to exploit the great advantages of tree-ring chronologies whilst avoiding some of the problems associated with climate reconstructions from physical ring properties (Helle and Schleser, 2004; McCarroll and Loader, 2004, 2005; Gagen et al., 2007).

However, the use of stable isotopes as dendroclimatic proxies is more recent than the use of physical tree-ring properties. The link between stable carbon isotope variations in plant compounds and temperature was first supposed by Urey (1947) who stated that "it seems probable that plant carbon compounds synthesised at coincidental temperatures may contain varying amounts of ^{13}C" (p. 579). Coincidentally, the first mass spectrometers for measuring light stable isotopes were developed at the same time (Nier, 1947; McKinney et al., 1950), so that Urey's provocative hypothesis could be tested. Although such pioneering studies in stable isotope research investigated the role and variability of carbon, oxygen and hydrogen isotopes in tree rings (Libby and Pandolfi, 1974; Libby et al., 1976; Epstein et al., 1976), the majority of studies were restricted to carbon isotopes, the stable isotope that is easiest to measure (Craig, 1953, 1954; Jansen, 1962; Wilson, 1977). One of the most important breakthroughs in the interpretation of stable isotope archives was the discovery of the mechanism of carbon isotopic fractionation in C_3 plants (Vogel, 1980), which provided the theoretical framework for most consecutive studies.

Meanwhile numerous studies have demonstrated the potential of stable isotopes in tree rings for quantitative terrestrial paleoclimate reconstructions (Switsur and Waterhouse, 1998; McCarroll and Loader, 2004, 2005; Robertson et al., 2008). While early studies oversimplified the climate-isotope relationship, treating trees as simple palaeothermometers recording, for example, the meteoric (source) water signal with little additional modification, it soon became clear that the tree-isotope archive is complex, containing physiological controls (e.g., a variable component reflecting stored or remobilized photosynthates) and ecological factors (e.g., disturbance, water and nutrient availability, disease, insect damage, and flowering). As a consequence, the isolation and extraction of a reliable and robust palaeoclimate signal from tree-ring isotope records remains challenging. Nevertheless, the processes and pathways through which the carbon, oxygen and hydrogen isotopes pass during assimilation and wood formation have been characterized (Vogel, 1980; Farquhar et al., 1982; Barbour and Farquhar, 2000; Roden et al., 2000). The resulting models (described below in Chapter 2.1 and 2.1) provide a robust framework for interpreting the isotopic composition of tree rings.

2.1 Stable isotope theory

Stable isotopes of carbon, oxygen and other elements are built into the ecosystem by physical, chemical and biological processes. Their relative abundances are unevenly distributed among and within different plant compounds as a result of biological-physicochemical processes during the uptake of CO_2 and H_2O, as well as during transformation processes within the plant. This phenomenon is called isotopic fractionation and is commonly expressed with the fractionation factor α:

$$\alpha = \frac{R_A}{R_B} \qquad (2.1)$$

It is defined as the ratio (R) of the rare-to-common (or heavy-to-light) abundance of any two isotopes in one chemical compound A (e.g., reactant) divided by the corresponding ratio of any other chemical compound B (e.g., product). Values of α are near one, therefore it is often more convenient to deal with Δ, which represents the isotopic discrimination

Chapter 2 Stable isotopes in dendroclimatology

$$\Delta = \alpha - 1 = \frac{R_\text{A}}{R_\text{B}} - 1 \qquad (2.2)$$

Processes leading to fractionation are influenced by environmental site conditions and produce distinctive stable isotopic ratio fingerprints in different organisms and substances, integrated in time and space.

Because of the very small absolute abundances of each isotope in any particular material (Ehleringer and Rundel, 1989), the stable isotope composition is expressed by convention relative to an international standard. Relative abundances can then be discussed more precisely than absolute isotope abundance ratios. This has led to the widely used "delta-notation" (McKinney et al., 1950), which expresses the isotope values as a ratio of heavy to light isotope (R_sample) in the delta notation as:

$$\delta^{xx}E = (\frac{R_\text{sample}}{R_\text{standard}} - 1) \times 1,000‰ \qquad (2.3)$$

where E is the Element of interest (e.g., C, O, H), xx is the atomic mass of the heavier isotope in the ratio and R is the absolute ratio of the element of interest (R_sample; e.g., $^{13}C/^{12}C$) of the element of interest relative to an international standard (R_standard; e.g., VPDB for carbon and VSMOW for oxygen[1]; Hayes, 1983). A positive δ value means therefore that the sample contains more of the heavier isotope than the given standard, a negative δ value means that the sample contains less of the heavier isotope than the standard.

The basis for carbon isotope variations in plants

Originally, carbon isotope variations in tree rings were analysed with the aim to compare them with the isotopic composition of atmospheric CO_2. Compared to atmospheric CO_2, the overall abundance of ^{13}C versus ^{12}C is lower in plant tissues, indicating that carbon isotope discrimination occurs during the incorporation of CO_2 into plant biomass. It was soon discovered that variations in the carbon isotope ratios in tree rings are not just reproducing changes in atmospheric CO_2, but are also indicating

---

[1]The Vienna Pee Dee Belemnite (VPDB) and Vienna Standard Mean Ocean Water (VSMOW) standards are provided by the International Atomic Energy Agency (IAEA), Vienna, Austria

2.1 Stable isotope theory

environmental changes and climate variations (Francey and Farquhar, 1982; Freyer and Belcay, 1983; Ehleringer and Rundel, 1989).

Trees assimilate atmospheric CO_2, which enters and diffuses out of the leaves through the stomata. As CO_2 diffuses through the stomata into the intercellular spaces, molecules of ^{12}C diffuse more readily into the leaf than ^{13}C, resulting in a "fractionation due to diffusion". This fractionation factor remains at about -4.4‰ (a, eqn. 2.4 and Fig. 2.1). It is due to the different mobility of isotopically heavy and light CO_2 and is largely independent of factors such as temperature or vapor pressure and even to all but the most extreme changes in stomatal aperture (≤ 0.1 μm) below which interactions between gas molecules and guard cells start to become important (Farquhar and Lloyd, 1993). Once inside the chloroplast, further fractionation occurs during photosynthesis, when CO_2 is combined enzymatically with leaf water to produce sugars via carboxylation. In C$_3$ plants, the photosynthetic enzyme Rubisco (Ribulose-1,5-bisphosphate carboxylase/oxygenase) tends to use ^{12}C in preference to ^{13}C, resulting in a "fractionation due to carboxylation" of about -27‰ (b, eqn. 2.4 and Fig. 2.1) (Farquhar et al., 1982; Francey and Farquhar, 1982; Ehleringer and Vogel, 1993). Plants performing another type of photosynthesis, C4 and CAM plants, show less isotopic fractionation than C3 plants due to a different photosynthetic enzyme (phosphoenolpyruvate–carboxylase).

The interplay between these fractionation factors and the resulting isotopic ratio of the carbon fixed by the plant ($\delta^{13}C_{plant}$) can be summarized in the following equation (Farquhar et al., 1982):

$$\delta^{13}C_{plant} = \delta^{13}C_{air} + a + (b - a) \times \left(\frac{c_i}{c_a}\right) \quad (2.4)$$

taking into account $\delta^{13}C_{air}$ of ambient CO_2, fractionation factors a and b, and the ratio of partial pressure of CO_2 inside (c_i) and outside (c_a) the leaf (compare Fig. 2.1). The fractionations occurring during diffusion and carboxylation are continuous and additive. If photosynthesis consumes CO_2 faster than it can be replaced by stomatal conductance, the relative pool of $^{12}CO_2$ would decrease, resulting in a decrease in fractionation against ^{13}C and an enriched $\delta^{13}C$ value. Any change in the rate of assimilation will therefore influence c_i. Similarly, a change in stomatal conductance will affect the rate at which the internal CO_2 can be replaced. If stomatal conductance is low and little CO_2 enters the leaf, c_i approaches 0 and the "fractionation due to diffusion" (a) more strongly determines the $\delta^{13}C_{\text{plant}}$. If stomatal conductance is high and CO_2 can exchange unhampered, $c_i \approx c_a$ so that $\delta^{13}C$ is more strongly determined by the "fractionation due to carboxylation" (b). In fact, the isotopic discrimination is situated somewhere in between these two extremes, slightly nearer the carboxylation-limited value, resulting in $\delta^{13}C_{\text{plant}}$ values between -22‰ and -35‰ for C_3 plants (C_4 plants vary between -10‰ and -18‰). If external factors, such as climate forcing, directly influence any of these controls, they will be integrated by the plant during photosynthesis. This can also be expressed as a change of instantaneous water use efficiency (WUE_i; Fig. 2.1) of the plant and the $\delta^{13}C$ of the product (Loader et al., 2007):

$$WUE_i = \frac{A}{g} = c_a \times [1 - \frac{c_i}{c_a}] \times (0.625) \qquad (2.5)$$

where A describes the photosynthetic rate and g the stomatal conductance.

Strictly speaking these formulae only consider fractionation processes occurring during photosynthesis and do not account for any additional fractionations associated with biochemical processes later on. Lipids or lignin for example can be depleted in $\delta^{13}C$ by as much as 3-6‰ relative to bulk organic matter whereas carbohydrates such as glucose and sucrose and related polymers such as starch and cellulose can be enriched by 1-3‰ (Ehleringer and Rundel, 1989; Flanagan et al., 2005). Nevertheless photosynthates formed in plant leaves build the basis of the $\delta^{13}C$ signature in the tree ring as they are transformed into cellulose. As long as only a constant shift is imprinted onto the cellulose, this does not influence the extraction of an environmental signal.

In summary, the dominant climatic controls on carbon isotopic fractionation are factors that control either stomatal conductance, such as relative humidity and soil moisture status, or the photosynthetic rate, such as light levels and leaf temperature (McCarroll and Loader, 2004; Loader et al., 2007). In addition, there are indirect climatic associations between these dominant forcing mechanisms and a range of quantifiable meteorological variables (including solar irradiance, cloud cover, temperature, precipitation amount, and relative humidity). Although not directly controlling the isotopic variations, these variables may correlate sufficiently well with both the forcing mechanisms and the tree-ring isotope signature, so that useful paleoclimatic information can be provided.

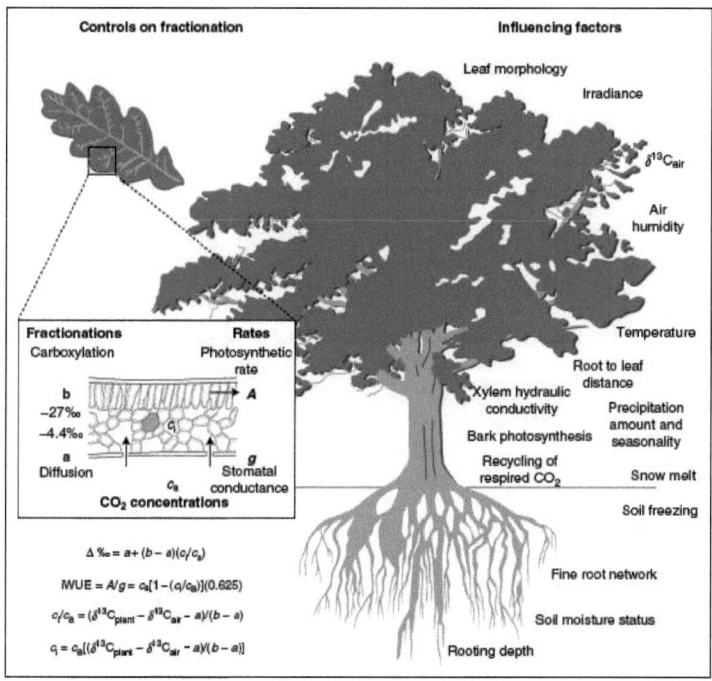

Figure 2.1: Diagram of a broad-leaf tree showing the main controls on the fractionation of carbon isotopes and the environmental factors that influence them. The equations are described in the text. [After McCarroll and Loader (2005: p. 74)]

The basis for oxygen isotope variations in plants

Oxygen and hydrogen isotopes enter the tree through the root system in the form of H_2O and move up the xylem to the leaves, where evaporation leads to preferential loss of the lighter isotope (compare Fig. 2.2). Oxygen and hydrogen isotope fractionation should be strongly related, while the discussion here is focussed on oxygen. The $\delta^{18}O$ composition of precipitation is related to local air mass characteristics and varies therefore at local and global scales depending on latitude, oceanic and continental regions, low and high elevation and temperature according to evaporation and condensation effects in air masses (Siegenthaler and Oeschger, 1980; Rozanski et al., 1993; Cole et al., 1999; Jouzel et al., 2000). No fractionation occurs during the water uptake of the trees, but as the water is taken from the soil and not directly from precipitation, fractionation and mixing can occur before the water is incorporated (Darling, 2004). Trees with shallow root systems may sample water dominated by local precipitation and evaporation regimes, whereas trees with deeper roots may access winter recharge or ground water, which both may potentially bear little resemblance to meteoric water characteristics from the growing season (Robertson et al., 2001; Loader et al., 2007).

Water taken up by the roots is transported through the xylem without any fractionation until it reaches the leaf. In the leaves an enrichment in ^{18}O occurs as $H_2{}^{16}O$ evaporates to the atmosphere more readily than $H_2{}^{18}O$ (Dongmann et al., 1974; Flanagan et al., 1991) amounting to as much as +20‰ (Saurer et al., 1998). The extent of ^{18}O enrichment of leaf water above xylem water at the sites of evaporation ($\Delta^{18}O_e$) is given by:

$$\Delta^{18}O_e = \epsilon^* + \epsilon_k + (\Delta^{18}O_v - \epsilon_k) \times \frac{e_a}{e_i} \qquad (2.6)$$

where ϵ^* is the proportional depression of water vapor pressure by the heavier water molecules, ϵ_k is the fractionation as water diffuses through the stomata and leaf boundary layer (kinetic fractionation factor), $\Delta^{18}O_v$ is the oxygen isotope composition of water vapour in the atmosphere (relative to source water) and the ratio of intercellular to atmosphere vapor pressures (e_a/e_i) (Craig and Gordon, 1965; Dongmann et al., 1974; Barbour et al., 2001, 2002).

This model accounts for the isotopic enrichment at the evaporation site, but overestimates leaf water enrichment as it treats leaf water as a single well-mixed pool

2.1 Stable isotope theory

and neglects any back-diffusion effects. Farquhar and Lloyd (1993) described a gradient within the leaf mesophyll generated by back-diffusion of isotopically enriched water from the evaporation surface into the leaf tissue, opposed by convection of non-enriched water from the xylem tissue to the evaporative surface. This so-called Péclet effect could cause isotopically lighter water than predicted by equation 2.6 (Yakir et al., 1990) and therefore could potentially modify the effective enrichment of sugars fixed prior to cellulose formation (Barbour and Farquhar, 2000; Barbour et al., 2001).

Sucrose formed in the leaf will therefore reflect the isotopic signature of leaf water to a certain extent, but with enrichment, as a result of isotopic exchange in intermediate products of photosynthesis with water, termed biochemical fractionation (Sternberg et al., 1986; Barbour et al., 2001). This $\delta^{18}O$ signal of leaf water imprinted on sucrose may partly be lost when sucrose is exchanging with less enriched xylem water during transportation and cellulose formation in the trunk (Sternberg et al., 1986; Hill et al., 1995; Roden et al., 2000). This post-photosynthesis exchange has been estimated to occur in as many as 40% of the oxygen atoms in the molecule. Thereby the amount of exchange depends on the rate of cellulose synthesis (= sucrose sink) as slow fluxes increase the rate of exchange (Barbour and Farquhar, 2000). On average, cellulose shows an enrichment of 27‰ compared to leaf water (Sternberg et al., 1986; Barbour et al., 2001).

Overall, $\delta^{18}O$ of tree-ring cellulose includes three variable components beyond source water variability: (1) the leaf water enrichment and the dampening of $\delta^{18}O$ at evaporation sites due to leaf water heterogeneity, (2) the biochemical fractionation during exchange with water, and (3) the $\delta^{18}O$ of water vapor outside the leaf (Roden et al., 2000). The oxygen isotopes reflect therefore the isotopic composition of source water (meteoric or ground water), which is related to temperature, and the degree of evaporative enrichment in the leaves, which is dominantly controlled by vapor pressure (air humidity). The relative strength of source water signal vs. leaf water enrichment is controlled by the degree of exchange with xylem water that occurs at the site of cellulose synthesis (Barbour and Farquhar, 2000; Loader et al., 2007). In a similar manner to carbon isotopes, oxygen isotopes may therefore preserve a past climate signal, induced through the isotopic composition of meteoric water (= source water) and an evaporative-transpirative signal in the leaves dominated by vapor pressure deficit, which are both indirectly linked to meteorological variables (temperature, precipitation amount, relative humidity, cloud cover).

Chapter 2 Stable isotopes in dendroclimatology

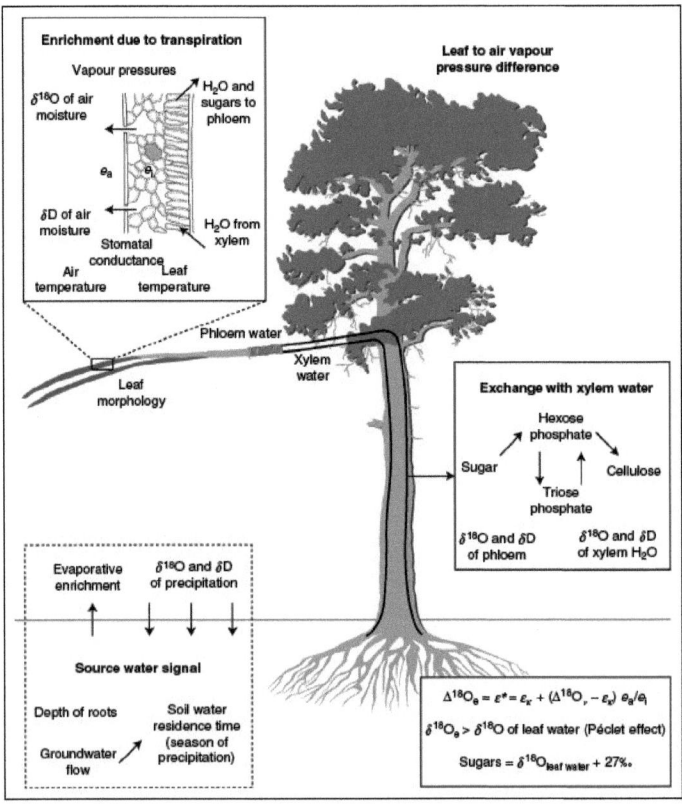

Figure 2.2: Diagram of a needle-leaf tree showing the main controls on the fractionation of the water isotopes and the environmental factors that influence them. The equations are described in the text. [Taken from McCarroll and Loader (2005: p. 73)]

2.2 State-of-the-art in stable isotope dendroclimatology

Although Baillie (1995) stated that "perhaps the pressure from environmental scientists will make stable isotopes as important in the next decade as radiocarbon was in the last" (p. 139), stable isotopes in tree rings didn't quite fulfill these ambitious expectations. Contrary to suggestions in early studies (e.g., Libby et al., 1976), none of the stable isotopes in tree rings provides a simple paleothermometer or is simply a measure of water availability. As the controls of isotopic fractionation are rather complex (see Chapter 2.1), none of the isotopes has a single controlling factor, and hence the dominant control will change according to the conditions under which the tree grew. Accordingly, variations in tree rings have been interpreted in many different ways, linking isotopic signatures with inferred palaeoclimatic signals including temperature, relative humidity, rainfall and the isotopic composition of the source water (McCarroll and Loader, 2004). However, there have been very few quantifiable, verifiable climatic reconstructions based upon stable isotope ratios from tree rings, so that the validity of the technique is questionable if stable isotopes merely contain the same climatic information as obtained from ring widths or wood density (Hughes, 2002). The widespread application of isotope dendroclimatology may have been constrained by laborious technical procedures. Recent progress in mass spectrometry (Saurer and Siegwolf, 2004; Filot et al., 2006; Leuenberger and Filot, 2007) and sample preparation techniques (Loader et al., 1997; Gaudinski et al., 2005; Rinne et al., 2005; Boettger et al., 2007) have revolutionized the field and enabled the routine preparation of stable isotope chronologies with a heightened degree of replication (Gagen et al., 2007) and temporal range (Robertson et al., 2008).

Sensitivity studies and paleoclimate reconstructions from carbon isotopes are manifold and cover a wide range of climate parameters. While Stewart et al. (1995) used $\delta^{13}C$ as a simple measure of water availability, Warren et al. (2001) compiled information from conifers worldwide and concluded that only in seasonally dry climates $\delta^{13}C$ is an indicator of drought stress and that variance in irradiance and nitrogen concentration can have as large an effect on isotope discrimination as water availability. Indeed, if tree growth is limited by moisture stress, strong correlations can be found between $\delta^{13}C$ and relative humidity and antecedent precipitation (e.g., Gagen et al.,

Chapter 2 Stable isotopes in dendroclimatology

2004). Leavitt and Long (1989), for example, reconstructed droughts from $\delta^{13}C$ for the southwestern United States. At locations where trees are rarely moisture-stressed, irradiance and temperature may be the climate factors that are strongly reflected in tree-ring $\delta^{13}C$ (e.g., Schleser et al., 1999; Hilasvuori et al., 2009). However, because under most conditions several factors control variability in $\delta^{13}C$, correlations with one single climate parameter are oversimplifications (McCarroll and Loader, 2004). Hot summers for example also tend to be dry so that temperature is correlated with relative humidity and antecedent precipitation, resulting in significant correlations between $\delta^{13}C$ and temperature and relative humidity (e.g., Edwards et al., 2000), temperature and precipitation (e.g., Kirdyanov et al., 2008), and all three climate parameters, respectively (e.g., Anderson et al., 1998; Treydte et al., 2001; McCarroll and Pawellek, 2001; Gagen et al., 2007. But while Gagen et al. (2007), with presumption of a stable relationship between sunshine duration and temperature, reconstructed summer temperature from $\delta^{13}C$ from *Pinus sylvestris* from northern Finland, a very recent study (Young et al., 2010) indicates summer cloud cover (i.e., summer sunshine duration) as the dominant driving force for $\delta^{13}C$ in *Pinus sylvestris* at tree-line sites in northern Norway. In summary, $\delta^{13}C$ signatures in tree rings strongly depend on local to regional site conditions and can contain a mixed climate signal. Paleoclimate reconstructions from $\delta^{13}C$ may therefore be of varying reliability and might be limited to sites which are strongly climate-limited (Reynolds-Henne et al., 2007).

In comparison to carbon isotopes there are fewer sensitivity studies and climate reconstructions based on well-replicated oxygen isotope series (Loader et al., 2007). For example, Saurer et al. (1997) demonstrated a close association between tree-ring $\delta^{18}O$ of beech (*Fagus sylvatica*) and the $\delta^{18}O$ composition of antecedent precipitation. Similarly, Robertson et al. (2001) were able to link oxygen isotopes from annually resolved oak latewood cellulose from east England with temperature, meteoric source water, and relative humidity, which supports the general structure of the mechanistic models (see Chapter 2.1). This work was continued by Saurer et al. (2002) who compared spatial and temporal trends in tree-ring $\delta^{18}O$ along the northern Eurasian tree line and successfully demonstrated the regional relevance and indicated the value of network approaches. One of the longest annually-resolved stable isotope records currently in existence is the 1000-year record from *Juniperus* growing in the Karakorum mountains in Pakistan presented by Treydte et al. (2006), which indicates a significant change in the precipitation amount during the 20th century.

2.2 State-of-the-art in stable isotope dendroclimatology

A very recent trend in isotope dendroclimatology is tree-ring isotope networks and multiproxy approaches, both of which seem to be promising, in particular for oxygen isotopes. The latter for instance has recently been demonstrated by Etien et al. (2008), who combined tree-ring-δ^{18}O from oak with grape harvest data from Fontainebleau, France, and built a growing season temperature reconstruction for the past 400 years, and by Loader et al. (2008), who indicated that a combination of standardized carbon and oxygen time-series into a single index increases the correlations with meteorological data in maritime climates. First results from tree-ring isotope networks reveal a strong coherence in the isotope records of different tree species across Europe, which is slightly more strongly expressed in the δ^{18}O than the δ^{13}C series (Treydte et al., 2007; Saurer et al., 2008). While Saurer et al. (2008) concluded that combining chronologies from different stands and different species enhances the potential for extending climate reconstructions into areas of temperate climate, Treydte et al. (2007) proposed to concentrate future reconstruction efforts on δ^{18}O in the European regions investigated in this study.

Finally, it is noteworthy that tree-ring stable isotopes are now being used to address a wide range of environmental issues. Miller et al. (2006) reconstructed the occurrence of tropical cyclones from δ^{18}O in longleaf pine tree rings, Liu et al. (2008) used tree-ring δ^{18}O to indicate the strength of Asian summer monsoon and Simard et al. (2008) detected insect defoliation events by applying tree-ring δ^{13}C and δ^{18}O. These environmentally broad applications of stable isotopes have previously not been thought possible and represent further valuable potential for plant-physiological and geophysical applications.

References

Anderson, W. T., Bernasconi, S. M., McKenzie, J. A., and Saurer, M. (1998). Oxygen and carbon isotopic record of climatic variability in tree ring cellulose (*Picea abies*): an example from central Switzerland (1913-1995). *Journal of Geophysical Research*, 103(D24):31625–31636; doi:10.1029/1998JD200040.

Baillie, M. G. L. (1995). *A Slice Through Time: Dendrochronology and Precision Dating*. Batsford, London.

Barbour, M. M., Andrews, T. J., and Farquhar, G. D. (2001). Correlations between oxygen isotope ratios of wood constituents of *Quercus* and *Pinus* samples from around the world. *Australian Journal of Plant Physiology*, 28(5):335–348; doi:10.1071/PP00083.

Barbour, M. M. and Farquhar, G. D. (2000). Relative humidity- and ABA-induced variations in carbon and oxygen isotope ratios of cotton leaves. *Plant, Cell and Environment*, 23:473–485; doi:10.1046/j.1365–3040.2000.00575.x.

Barbour, M. M., Walcroft, A. S., and Farquhar, G. D. (2002). Seasonal variation in $\delta^{13}C$ and $\delta^{18}O$ of cellulose from growth rings of *Pinus radiata*. *Plant, Cell and Environment*, 25(11):1483–1499; doi:10.1046/j.0016–8025.2002.00931.x.

Boettger, T., Haupt, M., Knoller, K., Weise, S. M., Waterhouse, J. S., Rinne, K. T., Loader, N. J., Sonninen, E., Jungner, H., Masson-Delmotte, V., Stievenard, M., Guillemin, M. T., Pierre, M., Pazdur, A., Leuenberger, M., Filot, M., Saurer, M., Reynolds, C. E., Helle, G., and Schleser, G. H. (2007). Wood cellulose preparation methods and mass spectrometric analyses of $\delta^{13}C$, $\delta^{18}O$ and nonexchangeable δ^2H values in cellulose, sugar, and starch: an interlaboratory comparison. *Analytical Chemistry*, 79(12):4603–4612; doi:10.1021/ac0700023.

Briffa, K. R. (2000). Annual climate variability in the Holocene: interpreting the message of ancient trees. *Quaternary Science Reviews*, 19:87–105; doi:10.1016/S0277–3791(99)00056–6.

Cole, J. E., Rind, D., Webb, R. S., Jouzel, J., and Healy, R. (1999). Climatic controls on interannual variability of precipitation $\delta^{18}O$: simulated influence of temperature, precipitation amount, and vapor source region. *Journal of Geophysical Research*, 104:14223–14235.

Cook, E. R., Briffa, K. R., Meko, D. M., Graybill, D. A., and Funkhouser, G. (1995). The segment length curse in long tree-ring chronology development for paleoclimatic studies. *The Holocene*, 5(2):229–237; doi:10.1177/095968369500500211.

Craig, H. (1953). The geochemistry of the stable carbon isotopes. *Geochimica et Cosmochimica Acta*, 3:53–92; doi:10.1016/0016-7037(53)90001-5.

Craig, H. (1954). Carbon-13 variations in *sequoia* rings and the atmosphere. *Science*, 119:141–143; doi:10.1126/science.119.3083.141.

Craig, H. and Gordon, L. (1965). Deuterium and oxygen^{18} variations in the ocean and marine atmospheres. In Tongoirgi, E., editor, *Proceedings of a Conference on Stable Isotopes in Oceanographic Studies and Palaeotemperatures*, pages 9–130, Pisa, Italy. Lischi and Figli.

Darling, W. G. (2004). Hydrological factors in the interpretation of stable isotope proxy data present and past: a European perspective. *Quaternary Science Reviews*, 23:743–770; doi:10.1016/j.quascirev.2003.06.016.

Dongmann, G., Nürnberg, H. W., Förstel, H., and Wagener, K. (1974). On the enrichment of $H_2^{18}O$ in leaves of transpiring plants. *Radiation, Environment and Biophysiology*, 11:41–52; doi:10.1007/BF01323099.

Edwards, T. W. D., Graf, W., Trimborn, P., Stichler, W., Lipp, J., and Payer, H. D. (2000). $\delta^{13}C$ response surface resolves humidity and temperature signals in trees. *Geochimica et Cosmochimica Acta*, 64(2):161–167; doi:10.1016/S0016-7037(99)00289-6.

Ehleringer, J. R. and Rundel, P. W. (1989). Stable isotopes: History, units and instrumentation. In Rundel, P. W., Ehleringer, J. R., and Nagy, K. A., editors, *Stable Isotopes in Ecological Research*, volume 68 of *Ecological Studies 68*. Springer, New York.

Ehleringer, J. R. and Vogel, J. C. (1993). Historical aspects of stable isotopes. In Ehleringer, J., Hall, A., and Farquhar, G., editors, *Stable Isotopes and Plant Carbon-Water Relations*, pages 9–18. Academic Press, New York.

Epstein, S., Yapp, C. J., and Hall, J. H. (1976). The determination of the D/H ratios of non-exchangeable hydrogen in cellulose extracted from aquatic and land plants.

Earth and Planetary Science Letters, 30:241–251; doi:10.1016/0012-821X(76)90251-X.

Eronen, M., Zetterberg, P., Briffa, K. R., Lindholm, M., Meriläinen, J., and Timonen, M. (2002). The supra-long Scots pine tree-ring record for Finnish Lapland: Part 1, chronology construction and initial inferences. *The Holocene*, 12(6):673–680; doi:10.1191/0959683602hl580rp.

Esper, J., Cook, E. R., and Schweingruber, F. H. (2002). Low-frequency signals in long tree-ring chronologies for reconstructing past temperature variability. *Science*, 295(5563):2250–2253; doi10.1126/science.1066208.

Esper, J., Wilson, R. J. S., Frank, D. C., Moberg, A., Wanner, H., and Luterbacher, J. (2005). Climate: past ranges and future changes. *Quaternary Science Reviews*, 24(20-21):2164–2166; doi:10.1016/j.quascirev.2005.07.001.

Etien, N., Daux, V., Masson-Delmotte, V., Stievenard, M., Bernard, V., Durost, S., Guillemin, M. T., Mestre, O., and Pierre, M. (2008). A bi-proxy reconstruction of Fontainebleau (France) growing season temperature from AD 1596 to 2000. *Climate of the Past*, 4(2):91–106.

Farquhar, G. D. and Lloyd, J. (1993). Carbon and oxygen isotope effects in the exchange of carbon dioxide between terrestrial plants and the atmosphere. In Ehleringer, J., Hall, A., and Farquhar, G., editors, *Stable Isotopes and Plant Carbon-Water Relations*, pages 47–70. Academic Press, New York.

Farquhar, G. D., O'Leary, M. H., and Berry, J. A. (1982). On the relationship between carbon isotope discrimination and the intercellular carbon dioxide concentration in leaves. *Australian Journal of Plant Physiology*, 9(2):121–137.

Filot, M., Leuenberger, M., Pazdur, A., and Boettger, T. (2006). Rapid on-line equilibration method to determine the D/H ratios of non-exchangeable hydrogen in cellulose. *Rapid Communications in Mass Spectrometry*, 20:3337–3344; doi:10.1002/rcm.274.

Flanagan, L. B., Comstock, J. P., and Ehleringer, J. R. (1991). Comparison of modeled and observed environmental influences on the stable oxygen and hydrogen isotope composition of leaf water in *Phaseolus vulgaris* L. *Plant Physiology*, 96:588–596; doi:10.1104/pp.96.2.588.

Flanagan, L. B., Ehleringer, J. R., and Pataki, D. E., editors (2005). *Stable isotopes and biosphere-atmosphere interactions: processes and biological controls*. Physiological Ecology. Elsevier Academic Press, London.

Francey, R. J. and Farquhar, G. D. (1982). An explanation of $^{13}C/^{12}C$ variations in tree rings. *Nature*, 297:28–31; doi:10.1038/297028a0.

Freyer, H. D. and Belcay, N. (1983). $^{13}C/^{12}C$ record in northern hemispheric trees during the past 500 years - anthropogenic impact and climatic superpositions. *Journal of Geophysical Research*, 88:6844–6852; doi:10.1029/JC088iC11p06844.

Fritts, H. C. (1976). *Tree Rings and Climate*. Academic Press, London, England.

Gagen, M., McCarroll, D., and Edouard, J.-L. (2004). Latewood width, maximum density, and stable carbon isotope ratios of pine as climate indicators in a dry subalpine environment, French Alps. *Arctic, Antarctic and Alpine Research*, 36(2):166–171; doi:.1657/1523–0430(2004)036[0166:LWMDAS]2.0.CO;2.

Gagen, M., McCarroll, D., Loader, N. J., Robertson, L., Jalkanen, R., and Anchukaitis, K. J. (2007). Exorcising the 'segment length curse': summer temperature reconstruction since AD 1640 using non-detrended stable carbon isotope ratios from pine trees in northern Finland. *The Holocene*, 17(4):435–446; doi:10.1177/0959683607077012.

Gaudinski, J. B., Dawson, T. E., Quideau, S., Schuur, E. A. G., Roden, J. S., Trumbore, S. E., Sandquist, D. R., Oh, S.-W., and Wasylishen, R. E. (2005). Comparative analysis of cellulose preparation techniques for use with ^{13}C, ^{14}C, and ^{18}O isotopic measurements. *Analytical Chemistry*, 77:7212–7224; doi:10.1021/ac050548u.

Grudd, H., Briffa, K. R., Karlén, W., Bartholin, T. S. Jones, P. D., and Kromer, B. (2002). A 7400-year tree-ring chronology in northern Swedish Lappland: natural climatic variability expresses on annual to millenial timescales. *The Holocene*, 12(6):657–665; doi:10.1191/0959683602hl578rp.

Hantemirov, R. M. and Shyatov, S. G. (2002). A continuous multimillennial ringwidth chronology in Yamal, northwestern Siberia. *The Holocene*, 12(6):717–726; doi:10.1191/0959683602hl585rp.

Hayes, J. M. (1983). Practice and principles of isotopic measurements in organic geochemistry. In Meinschein, W., editor, *Organic Geochemistry of Contemporaneous and Ancient Sediments*, pages 5–31. SEPM, Bloomington, Indiana.

Helle, G. and Schleser, G. H. (2004). Interpreting climate proxies from tree-rings. In Fischer, H., Floeser, G., Kumke, T., Lohmann, G., Miller, H., Negendank, J., and von Storch, H., editors, *The KIHZ project: Towards a synthesis of Holocene proxy data and climate models*, pages 129–148. Springer, Berlin.

Hilasvuori, E., Berninger, F., Sonninen, E., Tuomenvirta, H., and Jungner, H. (2009). Stability of climate signal in carbon and oxygen isotope records and ring width from Scots pine (*Pinus sylvestris* L.) in Finland. *Journal of Quaternary Science*, 24(5):469–480; doi:10.1002/jqs.1260.

Hill, S. A., Waterhouse, J. S., Field, E. M., Switsur, V. R., and Aprees, T. (1995). Rapid recycling of triose phosphates in oak stem tissue. *Plant, Cell and Environment*, 18(8):931–936; doi:10.1111/j.1365–3040.1995.tb00603.x.

Hughes, M. K. (2002). Dendrochronology in climatology - the state of the art. *Dendrochronologia*, 20(1-2):96–116; doi:10.1078/1125–7865–00011.

Jansen, H. S. (1962). Depletion of carbon^{13} in young kauri trees. *Nature*, 196(84-85; doi:10.1038/196084a0).

Jones, P. D. and Mann, M. E. (2004). Climate over past Millennia. *Reviews of Geophysics*, 42:RG2002; doi:10.1029/2003RG000143.

Jouzel, J., Hoffmann, G., Koster, R. D., and Masson, V. (2000). Water isotopes in precipitation: data/model comparison for present-day and past climates. *Quaternary Science Reviews*, 19(1-5):363–379; doi:10.1016/S0277–3791(99)00069–4.

Kirdyanov, A. V., Treydte, K. S., Nikolaev, A., Helle, G., and Schleser, G. H. (2008). Climate signals in tree-ring width, density and $\delta^{13}C$ from larches in Eastern Siberia (Russia). *Chemical Geology*, 252(1-2):31–41; doi:10.1016/j.chemgeo.2008.01.023.

Leavitt, S. W. and Long, A. (1989). Drought indicated in carbon^{13}/carbon^{12} ratios of southwestern tree rings. *Journal of the American Water Resources Association*, 25(2):341–347; doi:10.1111/j.1752–1688.1989.tb03070.x.

Leuenberger, M. and Filot, M. (2007). Temperature dependencies of high-temperature reduction on conversion products and their isotopic signatures. *Rapid Communications in Mass Spectrometry*, 21:1587–1598; doi:10.1002/rcm.2998.

Leuschner, H. H., Saas-Klassen, U., Jansma, E., Bailie, M. G. L., and Spurk, M. (2002). Subfossil European bog oaks: population dynamics and long-term growth

2.2 State-of-the-art in stable isotope dendroclimatology

depressions as indicators of changes in the Holocene hydro-regime and climate. *The Holocene*, 12(6):695–706; doi:10.1191/0959683602hl584rp.

Libby, L. M. and Pandolfi, L. J. (1974). Temperature dependence of isotope ratios in tree rings. *Proceedings of the National Academy of Sciences*, 71(6):2482–2486; doi:10.1073/pnas.71.6.2482.

Libby, L. M., Pandolfi, L. J., Payton, P. H., Marshall III, J., Becker, B., and Giertz-Sienbenlist, V. (1976). Isotopic tree thermometers. *Nature*, 261(284-288; doi:10.1038/261284a0).

Liu, Y., Cai, Q. F., Liu, W. G., Yang, Y. K., Sun, J. Y., Song, H. M., and Li, X. (2008). Monsoon precipitation variation recorded by tree-ring $\delta^{18}O$ in arid Northwest China since AD 1878. *Chemical Geology*, 252(1-2):56–61; doi:10.1016/j.chemgeo.2008.01.024.

Loader, N. J., McCarroll, D., Gagen, M., Robertson, I., and Jalkanen, R. (2007). Extracting climatic information from stable isotopes in tree rings. In Dawson, T. and Siegwolf, R., editors, *Stable Isotopes as Indicators of Ecological Change*, Terrestrial Ecology, pages 27–48. Elsevier Academic Press, San Diego.

Loader, N. J., Robertson, I., Barker, A. C., Switsur, V. R., and Waterhouse, J. S. (1997). An improved technique for the batch processing of small wholewood samples to α-cellulose. *Chemical Geology*, 136(3-4):313–317; doi:10.1016/S0009-2541(96)00133-7.

Loader, N. J., Santillo, P. M., Woodman-Ralph, J. P., Rolfe, J. E., Hall, M. A., Gagen, M., Robertson, I., Wilson, R., Froyd, C. A., and McCarroll, D. (2008). Multiple stable isotopes from oak trees in southwestern Scotland and the potential for stable isotope dendroclimatology in the maritime climatic regions. *Chemical Geology*, 252:62–71; doi:10.1016/j.chemgeo.2008.01.006.

McCarroll, D. and Loader, N. J. (2004). Stable isotopes in tree rings. *Quaternary Science Reviews*, 23(7-8):771–801; doi:10.1016/j.quascirev.2003.06.017.

McCarroll, D. and Loader, N. J. (2005). *Isotopes in Tree Rings*, volume 10 of *Developments in Paleoenvironmental Research Series*, chapter 2, pages 67–116. Springer.

McCarroll, D. and Pawellek, F. (2001). Stable carbon isotope ratios of *Pinus sylvestris* form northern Finland and the potential for extracting a climate signal from long Fennoscandian chronologies. *The Holocene*, 11(5):517–526; doi:10.1191/095968301680223477.

McKinney, C. R., McCrea, J. M., Epstein, S., Allen, H. A., and Urey, H. C. (1950). Improvements in mass spectrometers for the measurement of small differences in isotope bundance ratios. *The Review of Scientific Instuments*, 21(8):724; doi:10.1063/1.1745698.

Miller, D. L., Mora, C. I., Grissino-Mayer, H. D., Mock, C. J., Uhle, M. E., and Sharp, Z. (2006). Tree-ring isotope records of tropial cyclone activity. *Proceedings of the National Academy of Sciences*, 103:14294–14297; doi:10.1073pnas.0606549103.

Nier, A. O. (1947). A mass spectrometer for isotope and gas analysis. *The Review of Scientific Instuments*, 18(6):398–411.

Reynolds-Henne, C. E., Siegwolf, R. T. W., Treydte, K. S., Esper, J., Henne, S., and Saurer, M. (2007). Temporal stability of climate-isotope relationships in tree rings of oak and pine (Ticino, Switzerland). *Global Biogeochemical Cycles*, 21(4):GB4009; doi:10.1029/2007GB002945.

Rinne, K. T., Boettger, T., Loader, N. J., Robertson, I., Switsur, V. R., and Waterhouse, J. S. (2005). On the purification of α-cellulose from resinous wood for stable isotope (H, C and O) analysis. *Chemical Geology*, 222(1-2):75–82; doi:10.1016/j.chemgeo.2005.06.010.

Robertson, I., Leavitt, S. W., Loader, N. J., and Buhay, W. (2008). Progress in isotope dendroclimatology. *Chemical Geology*, 252(1-2):EX1–EX4; doi:10.1016/s0009-2541(08)00177-0.

Robertson, I., Waterhouse, J. S., Barker, A. C., Carter, A. H. C., and Switsur, V. R. (2001). Oxygen isotope ratios of oak in east England: implications for reconstructing the isotopic composition of precipitation. *Earth and Planetary Science Letters*, 191:21–31; doi:10.1016/S0012-821X(01)00399-5.

Roden, J. S., Lin, G., and Ehleringer, J. R. (2000). A mechanistic model for interpretation of hydrogen and oxygen isotope ratios in tree-ring cellulose. *Geochimica et Cosmochimica Acta*, 64(1):21–35; doi:10.1016/S0016-7037(99)00195-7.

Rozanski, K., Arguas-Arguas, L., and Gonfiantini, R. (1993). Isotopic patterns in modern global precipitation. In Swart, P., editor, *Climate Change in Continen-*

tal *Isotopic Records*, volume 78 of *Geophysical Monograph*, pages 1–36. American Geophysical Union, Washington, DC.

Saurer, M., Borella, S., and Leuenberger, M. (1997). $\delta^{18}O$ of tree rings of beech (*Fagus sylvatica*) as a record of $\delta^{18}O$ of the growing season precipitation. *Tellus B*, 49B:80–92; doi:10.1034/j.1600-0889.49.issue1.6.x.

Saurer, M., Cherubini, P., Reynolds-Henne, C. E., Treydte, K. S., Anderson, W. T., and Siegwolf, R. T. W. (2008). An investigation of the common signal in tree ring stable isotope chronologies at temperate sites. *Journal of Geophysical Research*, 113:G04035; doi:10.1029/2008JG000689.

Saurer, M., Robertson, I., Siegwolf, R., and Leuenberger, M. (1998). Oxygen isotope analysis of cellulose: an interlaboratory comparison. *Analytical Chemistry*, 70(10):2074–2080; doi:10.1021/ac971022f.

Saurer, M., Schweingruber, F., Vaganov, E. A., Shiyatov, S. G., and Siegwolf, R. (2002). Spatial and temporal oxygen isotope trends at the northern tree-line in Eurasia. *Geophysical Research Letters*, 29(9):10.1–10.4; doi:10.1029/2001GL013739.

Saurer, M., Siegenthaler, U., and Schweingruber, F. H. (1995). The climate-carbon isotope relationship in tree-rings and the significance of site conditions. *Tellus B*, 47(3):320–330; doi:10.1034/j.1600-0889.47.issue3.4.x.

Saurer, M. and Siegwolf, R. (2004). Pyrolysis techniques for oxygen isotope analysis of cellulose. In *Handbook of Stable Isotope Analytical Techniques*, volume 1, pages 497–508; doi:10.1016/B978-044451114-0/50025-9. Elsevier, New York.

Schleser, G. H., Helle, G., Lucke, A., and Vos, H. (1999). Isotope signals as climate proxies: the role of transfer functions in the study of terrestrial archives. *Quaternary Science Reviews*, 18(7):927–943; doi:10.1016/S0277-3791(99)00006-2.

Schweingruber, F. H. (1983). *Der Jahrring - Standort, Methodik, Zeit und Klima in der Dendrochronologie*. Verlag Paul Haupt, Bern and Stuttgart.

Schweingruber, F. H. (1996). *Tree Rings and Environment - Dendroecology*. Paul Haupt, Berne, Stuttgart, Vienna.

Siegenthaler, U. and Oeschger, H. (1980). Correlation of ^{18}O in precipitation with temperature and altitude. *Nature*, 285:314–317; doi:10.1038/285314a0.

Simard, S., Elhani, S., Morin, H., Krause, C., and Cherubini, P. (2008). Carbon and oxygen stable isotopes from tree-rings to identify spruce budworm outbreaks in the boreal forest of Québec. *Chemical Geology*, 252(1-2):80–87; doi:10.1016/j.chemgeo.2008.01.018.

Spurk, M., Leuschner, H. H., Bailie, M. G. L., Briffa, K. R., and Friedrich, M. (2002). Depositional frequency of German subfossil oaks: climatically and non-climatically induced fluctuations in the Holocene. *The Holocene*, 12(6):707–715; doi:10.1191/0959683602hl583rp.

Sternberg, L., DeNiro, M. J., and Savidge, R. (1986). Oxygen isotope exchange between metabolites and water during biochemical reactions leading to cellulose synthesis. *Plant Physiology*, 82:423–427; doi:10.1104/pp.82.2.423.

Stewart, G. R., Turnbull, M. H., Schmidt, S., and Erskine, P. D. (1995). ^{13}C natural abundance in plant communities along a rainfall gradient: a biological integrator of water availability. *Australian Journal of Plant Physiology*, 22(1):51–55; doi:10.1071/PP9950051.

Switsur, R. and Waterhouse, J. (1998). Stable isotopes in tree ring cellulose. In Griffiths, H., editor, *Stable Isotopes: Integration of Biological, Ecological and Geochemical Processes*, pages 303–321. BIOS Scientific Publishers, Oxford.

Treydte, K., Frank, D., Esper, J., Andreu, L., Bednarz, Z., Berninger, F., Boettger, T., D'Alessandro, C. M., Etien, N., Filot, M., Grabner, M., Guillemin, M. T., Gutierrez, E., Haupt, M., Helle, G., Hilasvuori, E., Jungner, H., Kalela-Brundin, M., Krapiec, M., Leuenberger, M., Loader, N. J., Masson-Delmotte, V., Pazdur, A., Pawelczyk, S., Pierre, M., Planells, O., Pukiene, R., Reynolds-Henne, C. E., Rinne, K. T., Saracino, A., Saurer, M., Sonninen, E., Stievenard, M., Switsur, V. R., Szczepanek, M., Szychowska-Krapiec, E., Todaro, L., Waterhouse, J. S., Weigl, M., and Schleser, G. H. (2007). Signal strength and climate calibration of a European tree-ring isotope network. *Geophysical Research Letters*, 34(24):L24302; doi:10.1029/2007GL031106.

Treydte, K. S., Schleser, G. H., Helle, G., Frank, D. C., Winiger, M., Haug, G. H., and Esper, J. (2006). The twentieth century was the wettest period in northern Pakistan over the past millennium. *Nature*, 440(7088):1179–1182; doi:10.1038/nature04743.

2.2 State-of-the-art in stable isotope dendroclimatology

Treydte, K. S., Schleser, G. H., Schweingruber, F. H., and Winiger, M. (2001). The climatic significance of δ^{13}C in subalpine spruces (Lötschental, Swiss Alps). *Tellus B*, 53(5):593–611; doi:10.1034/j.1600-0889.2001.530505.x.

Urey, H. C. (1947). The thermodynamic properties of isotope substances. *Journal of the Chemical Society of London*, 85:562–581; doi:10.1039/JR9470000562.

Vogel, J. C. (1980). Fractionation of the carbon isotopes during photosynthesis. *Sitzungsberichte der Heidelberger Akademie der Wissenschaften*, pages 111–135.

Warren, C. R., McGrath, J. F., and Adams, M. A. (2001). Water availability and carbon isotope discrimination in conifers. *Oecologia*, 127(4):476–486; doi:10.1007/s004420000609.

Wilson, A. T. (1977). ^{13}C/^{12}C in cellulose and lignin as paleothermometers. *Nature*, 265:133–135; doi:10.1038/265133a0.

Yakir, D., DeNiro, M. J., and Ephrath, J. E. (1990). Effects of water-stress on oxygen, hydrogen and carbon isotope ratios in two species of cotton plants. *Plant, Cell and Environment*, 13(9):949–955; doi:10.1111/j.1365-3040.1990.tb01985.x.

Young, G. H. F., McCarroll, D., Loader, N. J., and Kirchhefer, A. J. (2010). 500-years of summer near-ground solar radiation from stable carbon isotopes in Norwegian tree-rings. *The Holocene*, 20(3):315–324; doi:10.1177/0959683609351902.

3

Stable isotope coherence in the earlywood and latewood of tree-line conifers

Anne Kress[1,3]*, Giles H.F. Young[2]*, Matthias Saurer[1], Neil J. Loader[2], Rolf T.W. Siegwolf[1], and Danny McCarroll[2]

[1] *Laboratory of Atmospheric Chemistry, Paul Scherrer Institut, CH-5232 Villigen PSI, Switzerland*
[2] *Department of Geography, School of the Environment and Society, Swansea University, Singleton Park, Swansea SA2 8PP, UK*
[3] *Forest Ecology, Department of Environmental Sciences, Swiss Federal Institute of Technology Zurich, CH-8092 Zurich, Switzerland*

Published in *Chemical Geology* 268: 52–57; doi:10.1016/j.chemgeo.2009.07.008, 2009

*These authors contributed equally to this work

Chapter 3 Stable isotope coherence in the earlywood and latewood of tree-line conifers

Abstract

Annually resolved and replicated tree-ring stable isotope series have the potential to reconstruct growing season environmental parameters over multi-millennial timescales. As this archive may require only minimal statistical detrending, it has the potential to preserve a large portion of low frequency climate signals. To date, many studies have utilised only the latewood portion of the tree ring, in an attempt to minimize carry-over effects from previous year reserves and maximise the annual nature of the climate signal preserved. However, the old trees from tree-line locations, necessary to build long chronologies, often display narrow ring-widths (< 0.5 mm), making accurate earlywood-latewood separation difficult and particular time consuming. The resulting samples may also be too small for efficient cellulose purification or multiple isotopic determinations.

As photosynthates from the current year are predominantly used in conifer ring formation at marginal sites with short growing seasons, latewood separation may not be especially advantageous in determining a useful climate signal and therefore unnecessary where resources are limited. To test this hypothesis, Scots pine from Northern Norway and European larch from the Swiss Alps are used. Both sites are tree-line locations where growth is predominantly temperature limited. Tree rings were cut and extracted to cellulose for both the earlywood and latewood of each annual growth ring and stable carbon isotope ratios were measured.

Our results demonstrate a very high common carbon isotope signal between earlywood and latewood in both species ($r_{larch} = 0.68$ and $r_{pine} = 0.79$), which also show high correlations with summer temperature over the investigated period (AD 1980–2004 for larch and AD 1929–1978 for pine). High turnover rates and small reserve pools at these tree-line locations may account for these high common signals. These results suggest that for European tree-line conifers, the separation of earlywood from latewood is unnecessary to resolve an annual isotopic signal and make a reliable climate calibration. Using the whole ring may provide additional analytical advantages and consequently even improve climate calibrations.

Keywords European larch – Scots pine – carbon isotopes – dendroclimatology – Swiss Alps – Fennoscandia

3.1 Introduction

Stable carbon isotopes ($\delta^{13}C$) from tree rings are frequently used in environmental research, as they provide a continuous, annually resolved record of environmental conditions during growth (e.g., Anderson et al., 1998; McCarroll and Loader, 2004). Carbon isotopic ratios are controlled by the balance between stomatal conductance and photosynthetic rate (Farquhar et al., 1982; Leavitt and Long, 1988). In dry environments, the former is likely to dominate and stable carbon isotopes will correlate with air humidity and antecedent precipitation. In moist environments, where water stress is rare, the rate of photosynthesis is likely to dominate, which results in strong correlations with irradiance factors and growing season temperature (McCarroll et al., 2003; Loader et al., 2008).

Stable carbon isotopes may, therefore, offer potential for reconstructing long-term climatic information, as after a brief juvenile period, they appear to contain no long term age related trends (Gagen et al., 2007) and may therefore require only minimal statistical detrending to account for atmospheric changes in CO_2 during the industrial period (McCarroll et al., 2009). Although a topic of ongoing scientific debate, tree-ring stable carbon and oxygen isotopes are now being used to reconstruct lower frequency climate change (Treydte et al., 2006; Gagen et al., 2007). This is not to say that important lower-frequency environmental information cannot be extracted from physical proxies, a number of highly sophisticated well replicated millennial length reconstructions have been made using tree-ring widths and maximum latewood density, which include a high degree of low frequency variability (e.g., Büntgen et al., 2005, 2006; Grudd, 2008). It appears likely, therefore, that a combination of multiple proxies from tree rings may yield the best estimate of both low and high frequency climate variably (e.g., Hilasvuori et al., 2009).

To develop such isotope reconstructions, and to contribute to the current climate change debate, it is necessary to produce millennial length isotope tree-ring records with sufficient replication to robustly capture low-frequency trends, and produce statistically defined confidence intervals. This requires isotope measurements on a large scale, especially where annual resolution is required. Increasing isotope sample throughput, without compromising signal quality, is therefore highly desirable. One method of advancing this aim would be to utilize the whole ring (WR) rather than merely the latewood (LW) of individual tree rings. A problem with this approach is, however,

Chapter 3 Stable isotope coherence in the earlywood and latewood of tree-line conifers

that significant intra-annual variability has been recognised for carbon isotopes ratios (Loader et al., 1995; Helle and Schleser, 2004; Schulze et al., 2004; Li et al., 2005; Kagawa et al., 2006). This may be due to seasonal changes in micrometeorology (Leavitt, 1993; Livingston and Spittlehouse, 1996; Barbour et al., 2002; Schulze et al., 2004) or caused by the use of photosynthates from the previous year during the production of earlywood (EW), especially for deciduous hardwood species where the EW formation may commence prior to budburst (Pilcher, 1995). Carbon values in EW are, in this case, likely to be more associated with LW isotope values of the previous year rather than with LW of the current year (e.g., Hill et al., 1995; Switsur et al., 1995). Although some studies also find isotopically enriched EW in conifer species, others suggest that conifers do not seem to rely on stored carbon reserves (Dickmann and Kozlowski, 1970; Glerum, 1980; Barbour et al., 2002; Helle and Schleser, 2004). EW production would therefore use only the current years photosynthates, making the separation of EW from LW unnecessary for an annually resolved isotope signal.

Several authors suggest that the use of LW is preferable to produce climate reconstructions (e.g., Switsur et al., 1995; McCarroll and Loader, 2004). However, while some studies use only LW from both deciduous and conifer species for climate reconstructions (e.g., Gagen et al., 2007; Etien et al., 2008) others utilize LW of deciduous species but WR of conifer species (e.g., Treydte et al., 2007) or WR for deciduous and conifer species (e.g., Reynolds-Henne et al., 2007). Published guidance on the subject is often contradictory. For example, while Kagawa et al. (2006) emphasize the need of EW/LW-separation for climate reconstruction work with narrow boreal rings, Weigl et al. (2008) suggest that stable isotopes from all tree ring components (EW, LW and WR) are suitable as climate proxies.

In this study, we assess the relationship between the carbon isotopic signatures of EW and LW from tree rings of two conifer species at two European temperature-limited tree-line locations. Our aim is to determine whether (1) a sufficiently homogeneous annualised isotope signal can be obtained from WR ring cellulose, and (2) whether this signal correlates sufficiently well with climate parameters to allow palaeoclimate reconstruction.

3.2 Materials and methods

We considered two European conifer species from two different tree-line locations, one latitudinal and the other altitudinal (Figure 3.1). Scots pine (*Pinus sylvestris* L.) from Forfjord, in coastal northwestern Norway, is close to its northern growth limit; while European larch (*Larix decidua* Mill.) in the Lötschental, southwestern Switzerland, at ∼2100 m a.s.l., is located near the inner Alpine tree line. Both sites have short vegetation periods, which, combined with low summer temperatures, produce distinct but narrow annual growth rings (typically ∼0.5 mm). Details of the site characteristics are given in Kirchhefer (2001) and Kress et al. (2009). In selecting conifer species for this study, Scots pine was identified as it retains its needles for several years and provides typical conifer characteristics. In contrary, European larch is a deciduous conifer and therefore possesses some characteristics of deciduous broad-leaf trees that may be reflected in the isotopic composition of the earlywood and latewood during

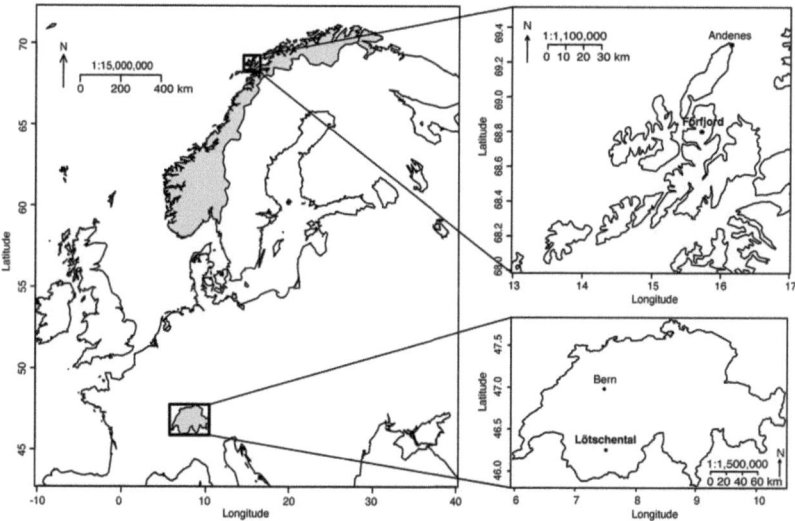

Figure 3.1: Scots pine (*Pinus sylvestris* L.) sampling site in costal Northern Norway, Forfjord (top right panel), and European larch (*Larix decidua* Mill.) sampling site in Southern Switzerland, Lötschental (bottom right panel).

57

ring formation. Both species are widely distributed and utilised in the production of millennial-long tree-ring series from tree-ring width (Kirchhefer, 2001; Büntgen et al., 2005, 2009) and maximum latewood density (Büntgen et al., 2006).

At both sites, mature trees were selected (one ∼ 300-year old specimen at the Norwegian site, two 250 to 300-year old specimens at the Swiss site) and cored at approximately 1.2 m using an increment borer. After surface preparation, tree-ring widths were measured (0.01 mm resolution) and cross-dated (Stokes and Smiley, 1968) against the local master chronologies (Kirchhefer, 2001; Büntgen et al., 2005). The program COFECHA Holmes (1983) was used to verify this cross-dating. The mean ring width was slightly larger for larch (0.92 mm) than for pine (0.68 mm) with a highly variable LW fraction of 17-25% in average for both species. All tree rings were separated into EW and LW before α-cellulose was extracted following standard procedures (Loader et al., 1997; Boettger et al., 2007). Cellulose samples of EW and LW were weighed for each year from AD 1929 to 1978 (Norwegian site) and AD 1980 to 2004 for two individual trees (Swiss site).

Carbon isotope ratios were measured online using a mass spectrometer interfaced with an elemental analyser, with samples combusted to CO_2 prior to mass spectrometric analysis. Analytical precision is typically ∼ 0.1‰ (σ_{n-1}, n = 10), with results expressed as $\delta^{13}C$, in per mille (‰) relative to the Vienna Pee Dee Belemnite standard (VPDB). All $\delta^{13}C$ values were corrected for the atmospheric $\delta^{13}C$ decrease due to fossil fuel burning since the beginning of the industrialisation (McCarroll and Loader, 2004).

For climate comparisons instrumental temperature data from Andenes (∼60 km distance from Forfjord; see Fig. 3.1) were used for the Norwegian samples (AD 1929–1978), while for the Swiss site instrumental temperature data from 19 Swiss stations (Auer et al., 2007) was applied for the period AD 1980 to 2004.

3.3 Results

The $\delta^{13}C$ results for EW and LW cellulose are presented in Table 3.1 and Figure 3.2, with comparison statistics shown in Table 3.2. Figure 3.2 demonstrates a highly significant agreement (p < 0.001) between the $\delta^{13}C$ values for EW and LW at both sites, with no major offsets throughout the series: while comparison statistics (Table

3.2) confirm the close similarity of EW and LW at both sites. The results from the Norwegian site (mean squared error – MSE of 0.11, an absolute difference – ABS of 0.28 and a coefficient of variance of R^2 of 0.79, $p < 0.001$) indicate a slightly closer agreement between EW and LW compared to the Swiss site. The Swiss results, however, display a high correlation between the two series ($R^2_{EW} = 0.57$; $R^2_{LW} = 0.76$) and the associated errors are low for both trees. When a mean of the two Swiss trees is taken, the comparison statistics improve (MSE = 0.20, ABS = 0.35 and $R^2 = 0.68$, $p < 0.001$). This is important as a mean of several trees is generally used when reconstructing climate and suggests that the addition of more trees may further reduce the difference between the mean EW and LW values. Moreover, the low squared correlation coefficients of current year EW with previous year LW of $R^2 = 0.03$ for the Swiss site and $R^2 = 0.08$ for the Norwegian site (Table 3.2) illustrate that the $\delta^{13}C$ signature of the current year EW is not significantly related to LW of the previous year.

Comparison with instrumental temperature data indicates the highest correlations between $\delta^{13}C$ and monthly temperature for the summer months at both sites. Only current year July, August, and July-August mean temperatures have significant correlations with $\delta^{13}C$ at both sites (at either $p < 0.001$ or 0.01, see Table 3.3). At the Norwegian site EW-$\delta^{13}C$ correlates more strongly with July than August temperature, while for LW it is vice versa. Both EW and LW correlate most strongly with July-August mean temperatures. A mean of the $\delta^{13}C$ values from EW and LW is taken, approximating the whole ring average, displays the highest overall correlation with temperature ($R^2 = 0.53$, July and August mean). Although the correlations are slightly weaker at the Swiss site, a similar pattern is displayed. The EW correlates most strongly with July while the LW correlates almost equally strong with July and August temperature. The highest correlation is also found between a whole ring (mean of EW and LW) $\delta^{13}C$ and July-August mean temperature ($R^2 = 0.44$). Although the simple averaging of EW and LW used for this climate comparison, rather than weighted averaging by relative mass, is an approximation of the total ring $\delta^{13}C$, the approach is supported by the research of Leavitt (2008), who shows pooled $\delta^{13}C$ results using weighted and non-weighted averaging to be nearly identical.

Chapter 3 Stable isotope coherence in the earlywood and latewood of tree-line conifers

Table 3.1: Earlywood (EW) and latewood (LW) $\delta^{13}C$ values of tree-ring cellulose obtained from one Scots pine tree (Tree1F) located at Forfjord in Norway and two European larch trees (Tree1L and Tree2L) from the Lötschental in Switzerland.

Norwegian $\delta^{13}C$ data (‰) (Forfjord)						Swiss $\delta^{13}C$ data (‰) (Lötschental)				
Year	EW Tree1F	LW Tree1F	Year	EW Tree1F	LW Tree1F	Year	EW Tree1L	LW Tree1L	EW Tree2L	LW Tree2L
1978	-24.36	-24.46	1953	-24.44	-24.40	2004	-22.61	-23.11	-23.23	-22.80
1977	-25.34	-25.03	1952	-25.01	-25.21	2003	-21.90	-21.21	-22.41	-21.66
1976	-25.60	-25.35	1951	-25.26	-24.95	2002	-23.30	-23.88	-24.20	-23.84
1975	-26.85	-26.13	1950	-24.83	-24.32	2001	-22.32	-22.20	-23.05	-22.59
1974	-25.66	-25.73	1949	-25.25	-25.68	2000	-22.63	-22.38	-23.43	-23.40
1973	-25.80	-26.11	1948	-24.76	-25.23	1999	-22.35	-23.24	-23.80	-23.70
1972	-24.60	-25.02	1947	-24.61	-25.13	1998	-22.12	-21.93	-23.44	-22.52
1971	-25.31	-25.50	1946	-23.64	-23.90	1997	-22.92	-23.40	-24.62	-24.04
1970	-24.28	-24.21	1945	-24.20	-24.25	1996	-22.71	-23.25	-24.07	-23.06
1969	-24.78	-24.38	1944	-24.54	-24.74	1995	-21.71	-22.79	-22.91	-22.64
1968	-26.45	-25.88	1943	-25.00	-25.17	1994	-21.69	-22.43	-23.14	-22.81
1967	-25.65	-25.86	1942	-24.79	-25.02	1993	-22.82	-22.66	-23.99	-23.76
1966	-25.35	-25.16	1941	-24.41	-24.26	1992	-22.75	-22.68	-23.56	-23.56
1965	-26.32	-25.46	1940	-25.06	-25.19	1991	-21.70	-21.08	-22.93	-21.82
1964	-25.15	-25.20	1939	-25.39	-25.14	1990	-21.87	-21.57	-23.52	-21.84
1963	-24.98	-24.65	1938	-25.01	-24.98	1989	-22.15	-22.53	-23.92	-23.37
1962	-26.07	-25.41	1937	-23.57	-23.78	1988	-22.13	-22.27	-24.02	-23.21
1961	-25.18	-25.35	1936	-24.27	-24.16	1987	-22.69	-22.56	-23.83	-23.15
1960	-24.97	-24.79	1935	-25.27	-25.12	1986	-22.05	-22.22	-22.42	-22.81
1959	-25.84	-25.77	1934	-23.38	-22.79	1985	-21.76	-21.70	-22.67	-22.27
1958	-25.51	-25.26	1933	-24.71	-24.79	1984	-22.08	-22.75	-23.53	-23.66
1957	-24.63	-24.77	1932	-25.46	-25.00	1983	-21.01	-20.34	-22.02	-21.29
1956	-24.89	-24.65	1931	-25.15	-24.78	1982	-22.00	-22.92	-23.22	-23.93
1955	-25.70	-25.58	1930	-24.34	-23.85	1981	-23.11	-22.96	-23.69	-23.37
1954	-24.15	-23.97	1929	-25.70	-25.32	1980	-22.09	-22.45	-23.02	-22.74

All values were corrected for atmospheric $\delta^{13}C$ changes after McCarroll and Loader (2004).

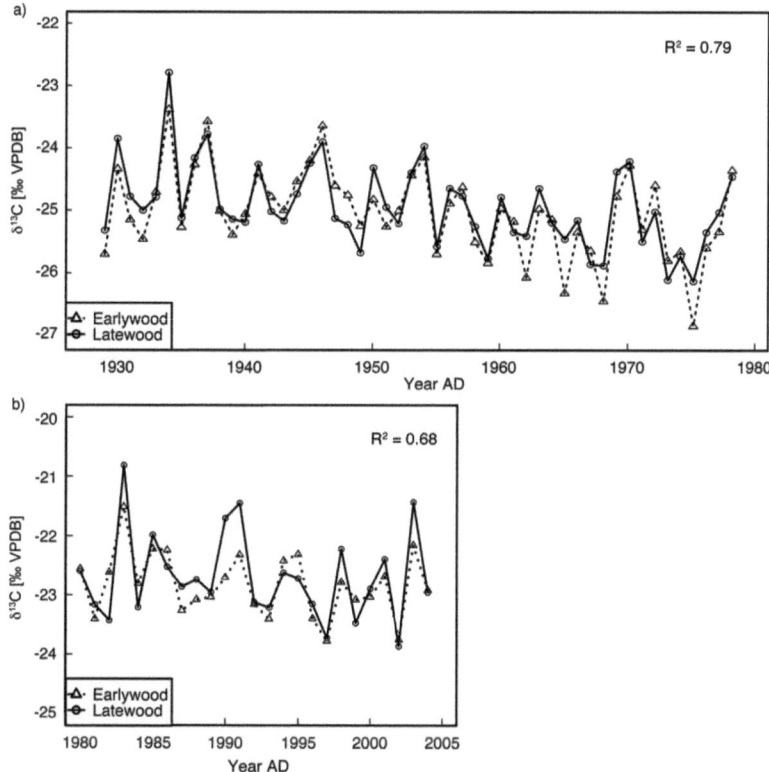

Figure 3.2: Earlywood and latewood intercomparison of $\delta^{13}C$ of tree-ring cellulose obtained from Scots pine, Forfjord, Norway (a) and European larch Mill.), Lötschental, Switzerland (b). All $\delta^{13}C$ values were corrected for the atmospheric changes after McCarroll and Loader (2004). The R^2 values are indicating coefficients of variance (Pearson's R^2) between earlywood and latewood over the entire 50-year-period at the Norwegian site and the 25-year period at the Swiss site.

Chapter 3 Stable isotope coherence in the earlywood and latewood of tree-line conifers

Table 3.2: Intercomparison statistics of earlywood (EW) and latewood (LW) $\delta^{13}C$ from one specimen of Scots pine from Forfjord (F), Norway and $\delta^{13}C$ from two specimens (and a mean of the two) of European larch, Lötschental (L), Switzerland (***p < 0.001).

Site	Tree No.	Mean Squared Error (MSE)	Mean Absolute Difference (ABS)	R^2 ($EW_{current}$ vs. $LW_{current}$)	R^2 ($EW_{current}$ vs. $LW_{previous}$)
Forfjord (Norway)	Tree1F	0.11	0.28	0.79***	0.08
Lötschental (Switzerland)	Tree1L	0.27	0.53	0.60***	0.01
	Tree2L	0.42	0.43	0.57***	0.03
	Average	0.20	0.35	0.68***	0.03

Table 3.3: Coefficient of variance (Pearson's R^2) between $\delta^{13}C$ values (‰ VPDB) corrected for atmospheric $\delta^{13}C$ changes after McCarroll and Loader (2004) for α-cellulose from tree-ring earlywood (EW), latewood (LW) and a mean of EW and LW and July, August and the mean July-August temperature from the nearest meteorological station at Andenes for the Norwegian site and the Swiss part of the HISTALP dataset for the Swiss site (***p < 0.001; **p < 0.01; *p < 0.05).

Site	Tree ring component	No. of years (n)	July (R^2)	August (R^2)	July-August (R^2)
Forfjord (Norway)	$\delta^{13}C$ EW	50	0.43***	0.33***	0.48***
	$\delta^{13}C$ LW	50	0.35***	0.51***	0.52***
	$\delta^{13}C$ EW-LW$_{mean}$	50	0.41***	0.43***	0.53***
Lötschental (Switzerland)	$\delta^{13}C$ EW	25	0.42***	0.10	0.37***
	$\delta^{13}C$ LW	25	0.29**	0.27**	0.42***
	$\delta^{13}C$ EW-LW$_{mean}$	25	0.37**	0.20*	0.44***

EW and LW are averaged over the two trees prior to calculation at the Swiss site.

3.4 Discussion and conclusions

The results from northwestern Norway and the Swiss Alps demonstrate a surprisingly strong coherence in the isotopic signal of EW and LW, considering that these two tree-ring components are formed during different periods of the growing season. Both the Norwegian and Swiss samples show high correlations and low mean squared error (MSE) and absolute difference (ABS) for EW and LW. Although the correlation coefficients between carbon isotope ratios in EW and LW are somewhat stronger at the Norwegian than at the Swiss site, this may be due to the correlation coefficients being calculated over periods of differing length (50 rather than 25 years) or caused by a slightly longer vegetation period in the Swiss Alps. Accordingly, slightly higher errors are found for the Swiss samples.

It has been demonstrated that EW may be influenced by the mobilisation of reserves at the beginning of the vegetation period, a phenomenon more associated with deciduous hardwood species (e.g., Loader et al., 1995; Helle and Schleser, 2004; Kagawa et al., 2006), than conifer species (Dickmann and Kozlowski, 1970; Glerum, 1980; Barbour et al., 2002). Helle and Schleser (2004) and Barbour et al. (2002) demonstrated that growth of pine species is not dependent on stored carbon during spring, resulting in an EW production, which is almost entirely related to current photosynthate. Cellulose laid down in the current spring will, therefore, reflect current conditions. This has previously been demonstrated by Dickmann and Kozlowski (1970) for *Pinus resinosa* from a forest nursery in Wisconsin, USA and by Glerum (1980) for *Pinus banksiana* in temperate climates of New Zealand. Previous research on $\delta^{13}C$ in EW and LW carried out on less temperature-limited sites is somewhat ambiguous: Weigl et al. (2008) show for middle European Sessile oak (*Quercus petraea* (Matt.) Liebl.) a significant correlation between $\delta^{13}C$ in EW and LW in the same year ($R^2 = 0.40$, $p < 0.001$, n = 50), but at the same time demonstrate higher correlations between LW and WR than EW and WR, concluding that the separation of LW from EW may be advisable. Jäggi et al. (2002), studying Norway spruce (*Picea abies*) on the Swiss Plateau, also demonstrate a high correlation between $\delta^{13}C$ in EW and LW of the current growth year ($R^2 = 0.76$, $p < 0.001$, n = 10), but point out that $\delta^{13}C$ in EW has a higher correlation with $\delta^{13}C$ in starch from the previous year's needles than with those of the current year. They therefore indicate that EW is more strongly influenced by biochemical fractionation (e.g., during starch formation) than by climatic conditions, whereas the latter are reflected in the isotopic signal of the LW. The site studied by them is not a tree-line location, situated in the Swiss plateau, it has a significantly longer vegetation period than northern or alpine tree-line sites and tree-growth is therefore not temperature- or moisture-limited. For the driest central part of the Swiss Rhone valley, where the major growth-limiting factor is moisture availability, Eilmann et al. (submitted for publication) found a high coherence between δ^{13}C in EW and LW of Scots pine ($R^2 = 0.81$, $p < 0.001$, $n = 51$). In our study, European larch, although deciduous, bears a striking resemblance between δ^{13}C in EW and LW. The relationship of δ^{13}C in EW and LW of "deciduous" larch is almost as strong as in "evergreen" pine. It could therefore be that at climatically limited sites, such as those in our study, the close isotopic similarity in EW and LW is rather explained by the prevailing site conditions and internal carbon budget than by the tree species alone.

Small disparities between the δ^{13}C signature in EW and LW can also be caused by micrometeorological changes during the current growing season (Livingston and Spittlehouse, 1996; Barbour et al., 2002; Leavitt, 2002; Li et al., 2005; Schulze et al., 2004) or by a stronger influence of biochemical fractionation (e.g., during starch formation) on the EW formation than on the LW production (Jäggi et al., 2002). As both species investigated in this study grow under highly temperature-limited conditions, we assume that high turnover rates and small reserve pools may account for such high common signals between EW and LW, and therefore prevent the mobilisation of any previous year reserves. Indeed, if these trees were using carbon from the previous year, a positive correlation between the LW of 1 year and the EW of the following year might be expected. No significant positive correlations of this kind were discovered. Additionally, both species under examination are known to display narrow ring-widths in mature specimens (Kirchhefer, 2001; Büntgen et al., 2005), making the accurate separation of EW from LW frequently challenging. LW generally produces rather small quantities of wood, which are more difficult to purify to cellulose and can result in conversion losses of up to 70%. This in turn may cause problems as too little sample material can result in inaccuracy of the isotopic measurements through blank effects, sample size variability or limited scope for replication that could impact the reliability of the resulting climate reconstructions. The use of the WR, therefore, has the advantage of providing a larger quantity of annually resolved material, which can be vital if multiple or repeat isotopic measurements are required, in particular for single-tree analysis. The small divergences between δ^{13}C in EW and LW found in this study may, in small part, be

3.4 Discussion and conclusions

due to biochemical effects and remobilised phytosynthates. However, factors such as: inaccurate separation of EW and LW, analytical uncertainty in isotopic measurements, or microclimatological changes through the growing season are likely to represent, at least, equivalent sources of uncertainty. However, at less marginal sites, where ring widths are generally wider, the role of stored photosynthates and biochemical effects may be more important. At such locations it may be advisable to perform a test-study to quantify the EW-LW relationship in δ^{13}C before embarking upon the development of a chronology extending several centuries.

At both sites we recognize that temperature is unlikely to represent the single dominant direct control on isotopic fractionation. However, in the absence of direct records of photon flux or vapour pressure deficit we rely upon the relationship observed between temperature and more directly associated climatic parameters such as sunshine duration and relative humidity to determine the relationship between climate and carbon isotopic fractionation. This limitation is acceptable if we assume the relationship between these climatic parameters to be stable through time and interpret these findings with an understanding of the mechanistic nature of carbon isotopic fractionation (Loader et al., 2008). Highest overall climate correlations were found when a mean of EW and LW was compared to a mean of July-August temperatures. Taking the average of EW and LW should almost be equivalent to an analysis of the whole ring, although this simple calculation neglects the usually different widths and therefore amounts of EW and LW. The slightly weaker climate correlations at the Swiss site compared to the Norwegian site, especially in the LW may be due to the relative shortness of the EW-LW series (25 years). A 105-year WR cellulose δ^{13}C series obtained from the same site displays a very high correlation with summer temperatures (Kress et al., 2009). However, our results suggest that the use of WR may account for all changes within the short growing season and provides in turn the most robust climate signal.

In summary, our results are not only of practical relevance for more efficient sample preparation, but may also contribute to the understanding of carbon allocation in conifers at tree-line locations. We conclude that the time-consuming and potentially inexact separation of LW from EW may not be necessary to obtain a homogeneous carbon isotope signal and accurate temperature calibration from conifers growing under temperature-limited conditions. The use of WR may even provide a better estimate of growing season temperature than either LW or EW alone.

Chapter 3 Stable isotope coherence in the earlywood and latewood of tree-line conifers

Acknowledgements

This work was funded by the EU project FP6-2004-GLOBAL-017008-2 (MILLENNIUM). Many thanks to the WSL dendro-unit for support during sampling and to L. Läubli for preparation and assistance with the measurements of the Swiss samples. Many thanks also to A. Kirchhefer for support during fieldwork at the Norwegian site. NJL, DMcC and GHFY thank the UK Royal Society, NERC NE/B501504/1, NE/C511805/1 and NER/S/A/2004/12466 for additional support. We acknowledge the constructive comments from two reviewers.

References

Anderson, W. T., Bernasconi, S. M., McKenzie, J. A., and Saurer, M. (1998). Oxygen and carbon isotopic record of climatic variability in tree ring cellulose (*Picea abies*): an example from central Switzerland (1913-1995). *Journal of Geophysical Research*, 103(D24):31625–31636; doi:10.1029/1998JD200040.

Auer, I., Böhm, R., Jurkovic, A., Lipa, W., Orlik, A., Potzmann, R., Schoner, W., Ungersbock, M., Matulla, C., Briffa, K., Jones, P., Efthymiadis, D., Brunetti, M., Nanni, T., Maugeri, M., Mercalli, L., Mestre, O., Moisselin, J. M., Begert, M., Muller-Westermeier, G., Kveton, V., Bochnicek, O., Stastny, P., Lapin, M., Szalai, S., Szentimrey, T., Cegnar, T., Dolinar, M., Gajic-Capka, M., Zaninovic, K., Majstorovic, Z., and Nieplova, E. (2007). HISTALP - historical instrumental climatological surface time series of the Greater Alpine Region. *International Journal of Climatology*, 27(1):17–46; doi:10.1002/joe.1377.

Barbour, M. M., Walcroft, A. S., and Farquhar, G. D. (2002). Seasonal variation in $\delta^{13}C$ and $\delta^{18}O$ of cellulose from growth rings of *Pinus radiata*. *Plant, Cell and Environment*, 25(11):1483–1499; doi:10.1046/j.0016-8025.2002.00931.x.

Boettger, T., Haupt, M., Knoller, K., Weise, S. M., Waterhouse, J. S., Rinne, K. T., Loader, N. J., Sonninen, E., Jungner, H., Masson-Delmotte, V., Stievenard, M., Guillemin, M. T., Pierre, M., Pazdur, A., Leuenberger, M., Filot, M., Saurer, M., Reynolds, C. E., Helle, G., and Schleser, G. H. (2007). Wood cellulose preparation methods and mass spectrometric analyses of $\delta^{13}C$, $\delta^{18}O$ and nonexchangeable $\delta^{2}H$

values in cellulose, sugar, and starch: an interlaboratory comparison. *Analytical Chemistry*, 79(12):4603–4612; doi:10.1021/ac0700023.

Büntgen, U., Esper, J., Frank, D. C., Nicolussi, K., and Schmidhalter, M. (2005). A 1052-year tree-ring proxy for Alpine summer temperatures. *Climate Dynamics*, 25(2-3):141–153; doi:10.1007/s00382-005-0028-1.

Büntgen, U., Frank, D., Carrer, M., Urbinati, C., and Esper, J. (2009). Improving Alpine summer temperature reconstructions by increasing sample size. In Kaczka, R., Malik, I., Owczarek, P., Gärtner, H., Helle, G., and Heinrich, I., editors, *TRACE - Tree Rings in Archaeology, Climatology and Ecology*, volume 7 of *Scientific Technical Report STR 09/03*, pages 36–43. GFZ Potsdam.

Büntgen, U., Frank, D. C., Niervergelt, D., and Esper, J. (2006). Summer temperature variations in the European Alps, AD 755-2004. *Journal of Climate*, 19(21):5606–5623; doi:10.1175/JCLI3917.1.

Dickmann, D. I. and Kozlowski, T. T. (1970). Mobilization and incorporation of photoassimilated ^{14}C by growing vegetative and reproductive tissues of adult *Pinus resinosa* Ait. trees. *Plant Physiology*, 45(3):284–288; doi:10.1104/pp.45.3.284.

Eilmann, B., Buchmann, N., Siegwolf, R., Saurer, M., Cherubini, P., and Rigling, A. (subm.). Inter- and intra-annual variation in tree-ring $\delta^{13}C$ indicating drought-induces carbon starvation in Scots pine. page submitted for publication.

Etien, N., Daux, V., Masson-Delmotte, V., Stievenard, M., Bernard, V., Durost, S., Guillemin, M. T., Mestre, O., and Pierre, M. (2008). A bi-proxy reconstruction of Fontainebleau (France) growing season temperature from AD 1596 to 2000. *Climate of the Past*, 4(2):91–106.

Farquhar, G. D., O'Leary, M. H., and Berry, J. A. (1982). On the relationship between carbon isotope discrimination and the intercellular carbon dioxide concentration in leaves. *Australian Journal of Plant Physiology*, 9(2):121–137.

Gagen, M., McCarroll, D., Loader, N. J., Robertson, L., Jalkanen, R., and Anchukaitis, K. J. (2007). Exorcising the 'segment length curse': summer temperature reconstruction since AD 1640 using non-detrended stable carbon isotope ratios from pine trees in northern Finland. *The Holocene*, 17(4):435–446; doi:10.1177/0959683607077012.

Glerum, C. (1980). Food sinks and food reserves of trees in temperate climates. *New Zealand Journal of Forestry Science*, 10(1):176–185.

Grudd, H. (2008). Tornetrask tree-ring width and density AD 500-2004: a test of climatic sensitivity and a new 1500-year reconstruction of north Fennoscandian summers. *Climate Dynamics*, 31(7-8):843–857; doi:10.1007/s00382-007-0358-2.

Helle, G. and Schleser, G. H. (2004). Beyond CO_2-fixation by Rubisco - an interpretation of $^{13}C/^{12}C$ variations in tree rings from novel intra-seasonal studies on broad-leaf trees. *Plant, Cell and Environment*, 27(3):367–380; doi:10.1111/j.0016-8025.2003.01159.x.

Hilasvuori, E., Berninger, F., Sonninen, E., Tuomenvirta, H., and Jungner, H. (2009). Stability of climate signal in carbon and oxygen isotope records and ring width from Scots pine (*Pinus sylvestris* L.) in Finland. *Journal of Quaternary Science*, 24(5):469–480; doi:10.1002/jqs.1260.

Hill, S. A., Waterhouse, J. S., Field, E. M., Switsur, V. R., and Aprees, T. (1995). Rapid recycling of triose phosphates in oak stem tissue. *Plant, Cell and Environment*, 18(8):931–936; doi:10.1111/j.1365-3040.1995.tb00603.x.

Holmes, R. L. (1983). Computer-assisted quality control in tree-ring dating and measurements. *Tree-Ring Bulletin*, 43:69–78.

Jäggi, M., Saurer, M., Fuhrer, J., and Siegwolf, R. (2002). The relationship between the stable carbon isotope composition of needle bulk material, starch, and tree rings in *Picea abies*. *Oecologia*, 131(3):325–332; doi:10.1007/s00442-002-0881-0.

Kagawa, A., Sugimoto, A., and Maximov, T. C. (2006). $^{13}CO_2$ pulse-labelling of photoassimilates reveals carbon allocation within and between tree rings. *Plant, Cell and Environment*, 29(8):1571–1584; doi:10.1111/j.1365-3040.2006.01533.x.

Kirchhefer, A. J. (2001). Reconstruction of summer temperatures from tree-rings of Scots pine (*Pinus sylvestris* L.) in coastal northern Norway. *The Holocene*, 11(1):41–52; doi:10.1191/095968301670181592.

Kress, A., Saurer, M., Büntgen, U., Treydte, K., Bugmann, H., and Siegwolf, R. T. W. (2009). Summer temperature dependency of larch budmoth outbreaks revealed by Alpine tree-ring isotope chronologies. *Oecologia*, 160(2):353–365; doi:10.1007/s00442-009-1290-4.

Leavitt, S. W. (1993). Seasonal ^{13}C/^{12}C changes in tree rings: species and site coherence, and a possible drought influence. *Canadian Journal of Forest Research*, 23:210–218; doi:10.1139/x93-028.

Leavitt, S. W. (2002). Prospects for reconstruction of seasonal environment from tree-ring δ^{13}C: baseline findings from the Great Lakes area, USA. *Chemical Geology*, 192(1-2):47–58; doi:10.1016/S0009-2541(02)00161-4.

Leavitt, S. W. (2008). Tree-ring isotopic pooling without regard to mass: no difference from averaging δ^{13}C values of each tree. *Chemical Geology*, 252(1-2):52–55; doi:10.1016/j.chemgeo.2008.01.014.

Leavitt, S. W. and Long, A. (1988). Stable carbon isotope chronologies from trees in the southwestern United States. *Global Biogeochemical Cycles*, 2(3):189–198; doi:10.1029/GB002i003p00189.

Li, Z. H., Leavitt, S. W., Mora, C. I., and Liu, R. M. (2005). Influence of earlywood-latewood size and isotope differences on long-term tree-ring δ^{13}C trends. *Chemical Geology*, 216(3-4):191–201; doi:10.1016/j.chemgeo.2004.11.007.

Livingston, N. J. and Spittlehouse, D. L. (1996). Carbon isotope fractionation in tree ring early and late wood in relation to intra-growing season water balance. *Plant, Cell and Environment*, 19(6):768–774; doi:10.1111/j.1365-3040.1996.tb00413.x.

Loader, N. J., Robertson, I., Barker, A. C., Switsur, V. R., and Waterhouse, J. S. (1997). An improved technique for the batch processing of small wholewood samples to α-cellulose. *Chemical Geology*, 136(3-4):313–317; doi:10.1016/S0009-2541(96)00133-7.

Loader, N. J., Santillo, P. M., Woodman-Ralph, J. P., Rolfe, J. E., Hall, M. A., Gagen, M., Robertson, I., Wilson, R., Froyd, C. A., and McCarroll, D. (2008). Multiple stable isotopes from oak trees in southwestern Scotland and the potential for stable isotope dendroclimatology in the maritime climatic regions. *Chemical Geology*, 252:62–71; doi:10.1016/j.chemgeo.2008.01.006.

Loader, N. J., Switsur, V. R., and Field, E. M. (1995). High-resolution stable isotope analysis of tree rings: implications of 'microdendroclimatology' for palaeoenvironmental research. *The Holocene*, 5(4):457–460; doi:10.1177/095968369500500408.

McCarroll, D., Gagen, M. H., Loader, N. J., Robertson, I., Anchukaitis, K. J., Los, S., Young, G. H. F., Jalkanen, R., Kirchhefer, A., and Waterhouse, J. S. (2009). Correction of tree ring stable carbon isotope chronologies for changes in the carbon dioxide content of the atmosphere. *Geochimica et Cosmochimica Acta*, 73(6):1539–1547; doi:10.1016/j.gca.2008.11.041.

McCarroll, D., Jalkanen, R., Hicks, S., Tuovinen, M., Gagen, M., Pawellek, F., Eckstein, D., Schmitt, U., Autio, J., and Heikkinen, O. (2003). Multiproxy dendroclimatology: a pilot study in northern Finland. *The Holocene*, 13(6):829–838; doi:10.1191/0959683603hl668rp.

McCarroll, D. and Loader, N. J. (2004). Stable isotopes in tree rings. *Quaternary Science Reviews*, 23(7-8):771–801; doi:10.1016/j.quascirev.2003.06.017.

Pilcher, J. R. (1995). Biological considerations in the interpretation of stable isotope ratios in oak tree-rings. In Frenzel, B., Stauffer, B., and Weiss, M., editors, *Paläoklimaforschung*, volume 15, pages 157–161. European Science Foundation, Strasbourg.

Reynolds-Henne, C. E., Siegwolf, R. T. W., Treydte, K. S., Esper, J., Henne, S., and Saurer, M. (2007). Temporal stability of climate-isotope relationships in tree rings of oak and pine (Ticino, Switzerland). *Global Biogeochemical Cycles*, 21(4):GB4009; doi:10.1029/2007GB002945.

Schulze, B., Wirth, C., Linke, P., Brand, W. A., Kuhlmann, I., Horna, V., and Schulze, E. D. (2004). Laser ablation-combustion-GC-IRMS - a new method for online analysis of intra-annual variation of $\delta^{13}C$ in tree rings. *Tree Physiology*, 24(11):1193–1201; doi:10.1093/treephys/24.11.1193.

Stokes, M. A. and Smiley, T. L. (1968). *An introduction to tree-ring dating.* (reprinted 1996). University of Arizona Press, Tucson, US, Chicago.

Switsur, V. R., Waterhouse, J. S., Field, E. M., Carter, A. H. C., and Loader, N. J. (1995). Stable isotope studies in tree rings from oak - techniques and some preliminary results. In Frenzel, B., Stauffer, B., and Weiss, M., editors, *Paläoklimaforschung*, volume 15, pages 129–140. European Science Foundation, Strasbourg.

Treydte, K., Frank, D., Esper, J., Andreu, L., Bednarz, Z., Berninger, F., Boettger, T., D'Alessandro, C. M., Etien, N., Filot, M., Grabner, M., Guillemin, M. T., Gutierrez, E., Haupt, M., Helle, G., Hilasvuori, E., Jungner, H., Kalela-Brundin, M., Krapiec, M., Leuenberger, M., Loader, N. J., Masson-Delmotte, V., Pazdur, A., Pawelczyk, S., Pierre, M., Planells, O., Pukiene, R., Reynolds-Henne, C. E., Rinne, K. T., Saracino, A., Saurer, M., Sonninen, E., Stievenard, M., Switsur, V. R., Szczepanek, M., Szychowska-Krapiec, E., Todaro, L., Waterhouse, J. S., Weigl, M., and Schleser, G. H. (2007). Signal strength and climate calibration of a European tree-ring isotope network. *Geophysical Research Letters*, 34(24):L24302; doi:10.1029/2007GL031106.

Treydte, K. S., Schleser, G. H., Helle, G., Frank, D. C., Winiger, M., Haug, G. H., and Esper, J. (2006). The twentieth century was the wettest period in northern Pakistan over the past millennium. *Nature*, 440(7088):1179–1182; doi:10.1038/nature04743.

Weigl, M., Grabner, M., Helle, G., Schleser, G. H., and Wimmer, R. (2008). Characteristics of radial growth and stable isotopes in a single oak tree to be used in climate studies. *Science of The Total Environment*, 393(1):154–161; doi:10.1016/j.scitotenv.2007.12.016.

4

Summer temperature dependency of larch budmoth outbreaks revealed by an Alpine stable isotope chronology

Anne Kress[1], Matthias Saurer[1], Ulf Büntgen[2], Kerstin S. Treydte[2], Harald Bugmann[3], and Rolf T.W. Siegwolf[1]

[1] *Laboratory of Atmospheric Chemistry, Paul Scherrer Institut, 5232 Villigen PSI, Switzerland*
[2] *Swiss Federal Research Institute WSL, 8903 Birmensdorf, Switzerland*
[3] *Forest Ecology, Department of Environmental Sciences, Swiss Federal Institute of Technology Zürich (ETH), 8092 Zürich, Switzerland*

Published in Oecologia 160: 353-365; doi:10.1007/s00442-009-1290-4, 2009

Chapter 4 Summer temperature dependency of larch budmoth outbreaks

Abstract Larch budmoth (LBM, *Zeiraphera diniana* Gn.) outbreaks cause discernable physical alteration of cell growth in tree rings of host subalpine larch (*Larix decidua* Mill.) in the European Alps. However, it is not clear if these outbreaks also impact isotopic signatures in tree-ring cellulose, thereby masking climatic signals. We compared LBM outbreak events in stable carbon and oxygen isotope chronologies of larch and their corresponding tree-ring widths from two high-elevation sites (1800–2200 m a.s.l.) in the Swiss Alps for the period AD 1900–2004 against isotope data obtained from non-host spruce (*Picea abies*). At each site, two age classes of tree individuals (150–250 and 450–550 years old) were sampled. Inclusion of the latter age class enabled one chronology to be extended back to AD 1650, and a comparison with long-term monthly resolved temperature data. Within the constraints of this local study, we found that: (1) isotopic ratios in tree rings of larch provide a strong and consistent climatic signal of temperature; (2) at all sites the isotope signatures were not disturbed by LBM outbreaks, as shown, for example, by exceptionally high significant correlations between non-host spruce and host larch chronologies; (3) below-average July to August temperatures and LBM defoliation events have been coupled for more than three centuries. Dampening of Alps-wide LBM cyclicity since the 1980s and the coincidence of recently absent cool summers in the European Alps reinforce the assumption of a strong coherence between summer temperatures and LBM defoliation events. Our results demonstrate that stable isotopes in tree-ring cellulose of larch are an excellent climate proxy enabling the analysis of climate-driven changes of LBM cycles in the long term.

Keywords Climate change – defoliation – dendrochronology – European larch – stable isotopes

4.1 Introduction

European larch (*Larix decidua* Mill.) is a species of trees growing at the highest altitudes in the European Alps and also one of the most temperature sensitive in this zone. With a longevity of 850+ years and its widespread utilizationas timber (Büntgen et al., 2006b), it is considered to be an ideal archive for summer temperature reconstructions. To date, however, only a few temperature reconstructions have been made from larch, and none of these have been based on stable isotopes. The only

4.1 Introduction

long-term temperature reconstructions for the greater Alpine region (GAR) were built from tree-ring width (TRW) and maximum latewood density (MXD) measurements of numerous living and historic larch trees (Büntgen et al., 2006a, 2005). Therefore, this species is vastly under-represented or even absent in tree-ring networks (Frank and Esper, 2005; Treydte et al., 2007). One possible factor for this scarcity of studies on larch may be related to earlier research (e.g., Schweingruber, 1985) that questioned the suitability of larch for climate reconstructions because of the periodic population waves of the grey larch budmoth (LBM, *Zeiraphera diniana* Gn.). The LBM is a foliage feeding Lepidopteran insect that is characterized by periodical outbreaks (8- to 10-year intervals), mainly in the interior valleys of the European Alps (Baltensweiler et al., 1977). The feeding of the LBM on larch needles causes massive defoliation that results in reduced tree growth and, thereby, interfere with the climatic signal contained in tree rings. While relatively short time series of these oscillations have been compiled from multi-annual surveys (Baltensweiler and Rubli, 1999), century-long series have been reconstructed from tree-ring data (Rolland et al., 2001; Weber, 1997) with a recent study spanning the past 1200 years (Esper et al., 2007). The population cycles of LBM are affected by numerous interactions with lower (e.g., host plants, prey) and/or higher trophic levels (e.g., predators, diseases; Berryman, 1996). Several mechanisms have been put forward to explain LBM oscillations, including behavioral changes in population quality (Baltensweiler, 1993a), budmoth-disease interactions (Anderson and May, 1980), induced host defenses (Fischlin, 1982), and host-parasitoid interactions (Turchin et al., 2003). However, it remains unclear how these cycles are modulated by climatic influences (Esper et al., 2007).

Periodic oscillations in abundance are one of the most remarkable characteristics of many animal population dynamics that are present in many forest insect species in the groups of Lepidoptera (Leavitt and Long, 1986; Simard et al., 2008; Weidner et al., 2006), Thysanoptera (Ellsworth et al., 1994), and Coleoptera (Haavik et al., 2008). At their peak abundance, populations may reach very high densities over large areas, resulting in massive defoliation (Berryman, 1996; Kurz et al., 2008; Mattson and Addy, 1975). It is unclear whether such events also alter isotopic ratios in tree-ring wood and cellulose.

Stable carbon ($\delta^{13}C$) and oxygen ($\delta^{18}O$) isotopes in tree-ring cellulose provide a continuous record of environmental conditions during tree growth (e.g., Anderson et al., 1998; McCarroll and Loader, 2004) that is complementary to classical dendrochronological variables such as ring width. Hence, isotopes have the potential to shed light on the relationship between population dynamics and climate variations as embodied in tree rings. Physical conditions and tree responses are reflected in both isotopes. $\delta^{13}C$ values depend on factors affecting photosynthetic uptake of CO_2 (such as light, relative humidity, temperature, atmospheric CO_2 concentration). They are therefore modulated by stomatal conductance and the rate of carboxylation during photosynthesis, which are regulated in turn not only by climatic variables but also by other factors, such as nutrient and water availability (Farquhar et al., 1982; Leavitt and Long, 1988). $\delta^{18}O$ values are constrained by the isotopic ratio of the source water (Roden et al., 2000), the isotopic signature of leaf water, as a consequence of evaporation and the Péclet effect (Barbour et al., 2004), biochemical fractionation during biosynthesis of photosynthetic sugars, and the re-equilibrium exchange between the carbohydrate and xylem water during tree-ring xylem cellulose synthesis (Yakir et al., 1990).

There have been very few studies carried out on the impact of phytophagous insects on the stable isotope composition in plant material and, in addition, the results of these are rather controversial. While some insect-isotope studies observed enriched $\delta^{13}C$ values in the tree rings of host species during infestations (Leavitt and Long, 1986; Simard et al., 2008) or hardly any effect (Weidner et al., 2006), others inferred a complete lack of infestation response based on unaltered $\delta^{13}C$ signatures and unaffected water use efficiency (the ratio between photosynthesis and transpiration) (Ellsworth et al., 1994; Haavik et al., 2008). Investigations of the influence of forest insect outbreaks on $\delta^{18}O$ values are extremely rare and restricted to Lepidoptera: while Weidner et al. (2006) described a slight decrease, Simard et al. (2008) observed no changes in $\delta^{18}O$ signatures during outbreak events. By combining enriched $\delta^{13}C$ and unaffected $\delta^{18}O$ values in outbreak years, Simard et al. (2008) concluded that there may be an influencing factor other than climate. If factors other than climate dominate the isotopic signatures during insect infestations, climate reconstructions will be biased. Disentangling the role of insect infestations and climate on isotopic signatures in tree-ring cellulose is thus essential to prevent artifacts and, consequently, errors in climate reconstructions.

We have performed a study in which we assessed the relationship between LBM outbreaks and the carbon and oxygen signatures obtained from tree-ring cellulose of larch (*Larix decidua* Mill.). We combined an intra-species comparison of larch from two different valleys within the main crest of the Swiss Alps, canton Valais (northern and southern aspect each; AD 1900–2004), with an inter-species comparison between the LBM host species larch and the LBM non-host species spruce (*Picea abies*) from one of these sites for tree individuals within an age class of 150–250 years. To extend our investigated period and to take into account possible age-dependent effects of tree individuals on isotopic signatures during LBM outbreaks and their response to climate, we built an additional isotope chronology from older individuals (450–550 years) extending back to AD 1660. Thus, our objectives are twofold: (1) to clarify whether LBM-induced defoliation events have an impact on isotopic signatures in tree-ring cellulose, thereby masking climate signals, and (2) to address the long-term relationship between climatic conditions and LBM oscillations.

4.2 Material and methods

4.2.1 Sampling strategy

Larch samples (*Larix decidua* Mill.) were collected at two locations situated within the subalpine belt of larch-Swiss Stone pine (*Pinus cembra*) forests: the Lötschental (LOE) and the Simplon (SIM) region, both near the main crest of the Swiss Alps, canton Valais (coordinates of the individual sites are provided in Table 4.1). The elevation of the sites varies between 1800 and 2200 m a.s.l. Sampling occurred within the framework of the EU project Millennium. In each valley, trees were cored twice at breast height using a 5-mm increment borer. Thereby, care was taken to select trees under similar growth conditions on north-facing (N) and south-facing (S) slopes near the upper tree line. At both sites and both aspects, trees within an age-class of 150–250 years (LOE-S, LOE-N, SIM-S, SIM-N) were sampled. In the Lötschental, additional samples from 450- to 550-year old individuals (CHRONO-N and CHRONO-S) were collected (Table 4.2).

4.2.2 Sample analysis

Tree-ring width was measured with a semi-automated RinnTech system (0.01-mm resolution; Heidelberg, Germany) and cross-dated following standard procedures (Stokes and Smiley, 1968) and the program COFECHA (Holmes, 1983) to ensure correct dating of each tree ring to its calendar year of formation. Four to five dominant trees from

Table 4.1: Site characteristics and detected gray larch budmoth (*Zeiraphera diniana* Gn.) events for each of the study sites

Site character	Study sites[a]					
	CHRONO-S	CHRONO-N	LOE-S	LOE-N	SIM-S	SIM-N
Latitude °N	46°26'	46°23'	46°26'	46°23'	46°12'	46°11'
Longitude °E	7°48'	7°47'	7°48'	7°47'	8°04'	8°03'
Aspect	140°	345°	140°	345°	220°	20°
Slope	35°	40°	35°	40°	25°	30°
Altitude	2100 m asl	2200 m asl	2100 m asl	2200 m asl	1800 m asl	1800 m asl
Time-period	1650-2004	1650-2004	1900-2004	1900-2004	1900-2004	1900-2004
No. of trees	3	2	4	4	4	4
Age class	400-500 yrs	400-500 yrs	150-250 yrs	150-250 yrs	150-250 yrs	150-250 yrs
LBM-Events 20th century	1981	1981	1981	1981	1981	1981
		1974		1974		
	1972		1972		1972	1972
	1963		1963	1963	1963	1963
	1954	1954	1954	1954	1954	1954
	1945	1945	1945	1945	1945	1945
	1937		1937		1937	1937
					1925	1925
	1923	1923	1923	1923		
	1915		1915			
		1912		1912		
	1908	1908	1908	1908	1908	1908

LBM, Gray larch budmoth (*Zeiraphera diniana* Gn.)

[a]Study sites: CHRONO-S, CHRONO-N, long-term chronology sites (S, south-facing; N, north-facing) in the Lötschental region; LOE-S, LOE-N, younger material from the Lötschental region; SIM-S, SIM-N, younger material from the Simplon region

4.2 Material and methods

Table 4.2: Long-term chronology of detected LBM events in 450- to 550-year-old individuals in the Lötschental region

LBM Events	19th Century	18th Century	17th Century
CHRONO	1896	1777	1685
(Average of S & N)	1888	1771	1675
	1880	1758	1668
	1856	1753	1660
	1830	1743	
	1821	1732	
	1813	1720	
	1801	1708	
		1703	

each field site were selected for isotope analyses. These trees were split year-by-year using a scalpel; earlywood and latewood were not separated because of the rings being too narrow, and there was therefore insufficient material for isotope measurements. Isotope analysis of 50 sub-samples showed a very strong correlation between latewood and earlywood isotope values, thus justifying whole-ring analysis (data not shown).

For the 20th century (AD 1900–2004), the CHRONO samples were measured on a single-tree basis (two cores per tree, two and three trees, respectively), while for the period AD 1899–1650, five trees and two cores per tree were pooled prior to analysis for each annual ring. The four trees (two cores per tree) of each additional stand (LOE-N, LOE-S, SIM-N, SIM-S) were treated with the same pooling approach (Tables 4.1, 4.2). The pooling of rings of the same year from different cores has been used successfully for climate analysis; this approach retains the annual resolution, but the workload for sample preparation is reduced (Treydte et al., 2007). Alpha cellulose was extracted using standard procedures (Boettger et al., 2007) adapted for larch (1 vol% $NaClO_2$ solution; only one NaOH step with 17 vol% NaOH at 25 °C), homogenized by sonification, and freeze-dried for 24 h. Carbon isotopic ratios were assigned with a reproducibility of 0.1‰ by combustion to CO_2 at 1025 °C in an elemental analyzer (EA-1110; Carlo Erba Thermoquest, Milan, Italy) coupled to an isotope ratio mass spectrometer (Delta S or Delta Plus XL; Thermo Finnigan Mat, Bremen, Germany).

Oxygen isotopic ratios were determined after pyrolysis to CO at 1080 °C, with a reproducibility of 0.3‰ using a continuous flow method similar to that used for carbon and with the same mass spectrometers connected via a variable open-split interface Conflo II (Thermo Finnigan Mat; Saurer and Siegwolf, 2004). The isotope values were expressed as a ratio of heavy to light isotope (R_{sample}) in the delta notation as

$$\delta^{13}C \text{ or } \delta^{18}O = (\frac{R_{sample}}{R_{standard}} - 1) \times 1,000‰ \qquad (4.1)$$

relative to an international standard ($R_{standard}$: VPDB for carbon and VSMOW for oxygen). All $\delta^{13}C$ isotope series were corrected for the decline in $\delta^{13}C$ of atmospheric CO_2 due to fossil fuel burning since the beginning of industrialization using ice core data supplemented with the isotope measurements of atmospheric CO_2 for recent years (Leuenberger, 2007).

To assess the strength of the isotopic signal, we calculated the mean inter-series correlations (RBAR) and the expressed population signal (EPS) for each isotope series. While RBAR is a measure of common variance between single series, EPS is an absolute measure of chronology variance (threshold usually set to EPS > 0.85; McCarroll and Loader, 2004).

Tree-ring width data were standardized using a 150-year cubic smoothing spline filter (Esper et al., 2002) to remove the age-trend while preserving mid-term variations. A single isotope and TRW chronology was developed for the CHRONO samples (AD 1650–2004; for the period AD 1900–2004, each tree separately) and all other sites (AD 1900–2004).

4.2.3 Climatic data

We used two alpine-wide meteorological datasets: (1) the HISTALP database (Auer et al., 2007) of monthly homogenized instrumental temperature (back to AD 1760) and precipitation measurements (back to AD 1800) for the GAR (4 − 19°E, 43 − 49°N, 0–3,500 m a.s.l.) and (2) the gridded dataset "European Alps Temperature and Precipitation reconstructions" (Casty et al., 2005) where instrumental data were combined with documentary proxy evidence at a 0.5° x 0.5° resolution covering the period AD 1500–2000. While the AD 1901–2000 period is based on CRU TS2 data

(Mitchell and Jones, 2005), the period prior to the 20th century is based on documentary evidence that does not include any tree-ring evidence at any time.

Annual isotopic data of combined N- and S-CHRONO samples (AD 1900–2004) were compared with all existing monthly temperature (19) and precipitation (32) series (averaged prior to calculation) obtained from the Swiss part of the HISTALP database. Correlation analysis was conducted on a monthly basis from March of the previous year to December of the current year as well as on seasonal averages, using bootstrapped correlation analysis for significance testing (P <0.05) (Guiot, 1991). To quantify temporal changes in the relationship between climate and tree-ring isotopes, we additionally calculated moving correlations over 40-year time windows, thus producing a time series of bootstrapped correlation coefficients on a monthly basis from January to October of the current year (Biondi and Waikul, 2004; Reynolds-Henne et al., 2007).

In addition, the monthly temperature reconstructions of Casty et al. (2005) from the grid cell 46.25°N 7.75°E covering the Lötschental were used for the analysis of the climate-isotope-LBM relationship over time (AD 1660–2000).

4.2.4 Grey larch budmoth and stable isotopes

To accurately detect LBM outbreaks, all samples were screened for particularly narrow TRW and/or irregular latewood cells (missing, malformed, or lighter brown) following the dendrochronological skeleton plot technique (Schweingruber et al., 1990) and compared to a TRW dataset of larch samples containing 78 samples from the Simplon region and 330 samples from the Lötschental region (Büntgen et al., 2005). Moreover, a dataset of 180 MXD series from the same sites (Büntgen et al., 2006b) was screened for exceptionally low MXD values caused by LBM outbreaks. The comparison of these data with historical records of LBM outbreaks reaching back to AD 1850 (Baltensweiler and Rubli, 1999) and back to AD 832 (Esper et al., 2007) enabled us to identify -on an annual basis- seven to nine LBM outbreaks within the 20th century and 21 over the 1650–1899 period. Some years were only detected on one of the slopes (Tables 4.1, 4.2).

Superposed epoch analysis (SEA) and Wilcoxon-Mann-Whitney tests were applied to identify potential isotope responses to outbreak events in the isotope records. The

Chapter 4 Summer temperature dependency of larch budmoth outbreaks

SEA method isolates signals that are difficult to detect against relatively large levels of background noise (Adams et al., 2003). We applied SEA to our LBM events as follows: for each detected defoliation event, an 11-year window was centered on the outbreak year (defined as year 0). This provides a reasonable interval for resolving responses to LBM outbreaks without having to extend from one LBM event into another (8- to 10-year interval). The Wilcoxon-Mann-Whitney test (also called Wilcoxon-Mann-Whitney rank sum test) was used to quantify differences between two independent groups. The P_{wilcox} value indicates whether there is a statistically significant difference between the medians of the two groups (this is the case $P_{wilcox} < 0.05$) (Stahel, 2002).

Isotopic values of host (larch) and non-host (spruce) species during LBM outbreak years were compared. To this end, we used a published carbon isotope (Treydte et al., 2001) and a new oxygen isotope dataset obtained from usually LBM-unaffected spruce (*Picea abies*). The sampled spruces originate from six mid- to high-elevation sites (1400–1950 m a.s.l.) of the northern and southern aspects in the Lötschental, Valais, Switzerland. They are located nearby our larch sampling sites and cover a period of 50 years (AD 1946–1995).

4.3 Results

4.3.1 Climate-isotope relationship (AD 1900-2004)

The separate analysis of five trees (two from the N-facing and three from the S-facing slope) from the CHRONO site for AD 1900–2004 showed a highly synchronous carbon and oxygen isotope signal in all trees, as indicated by the mean inter-series correlation (RBAR = 0.76 for carbon and 0.81 for oxygen) and the EPS (0.94 for carbon and 0.96 for oxygen). Both statistics thus suggest a strong climate forcing at this site. Absolute differences between the mean values of the trees were up to 1.6‰ for $\delta^{13}C$ and 1.5‰ for $\delta^{18}O$, but these are not considered further because all analyses were performed with the average site chronologies only.

Hence, we calculated correlations between the average $\delta^{13}C$- and $\delta^{18}O$-isotope series of old trees CHRONO (corrected for the decline in $\delta^{13}C$ of atmospheric CO_2) and monthly averaged temperature values (Fig. 4.1). Figure 4.1a shows the significant positive correlations between both isotope series and summer temperature ($P < 0.05$).

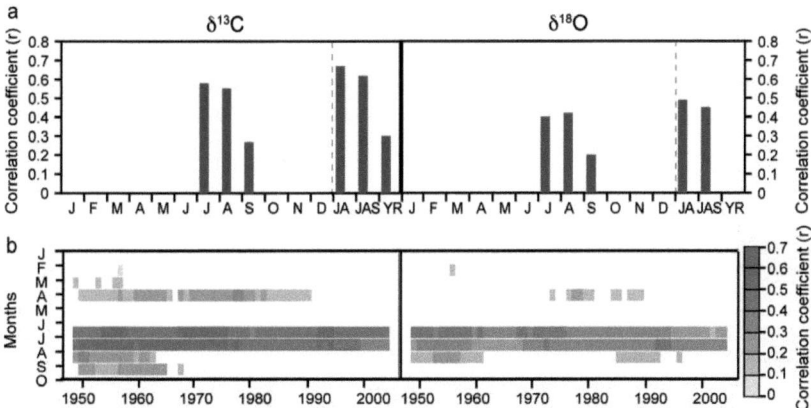

Figure 4.1: a) Bootstrapped significant correlation coefficients (P < 0.05) calculated for the stable carbon isotope (δ^{13}C; left) and stable oxygen isotope (δ^{18}O; right), averaged monthly temperature for January (J) to December (D) of the current year, averaged for July/August (JA), July/August/September (JAS) and for the entire year (YR) for AD 1900–2004 using the HISTALP dataset.

b) Forty-year running mean values of significant bootstrapped correlation coefficients (P < 0.05; insignificant correlations set to 0) for δ^{13}C (left) and δ^{18}O (right) and monthly temperature. As indicated on the y-axis, the calculation was performed for individual months January (J) to October (O) of the current year for AD 1900–2004 (40-year window determines 1940 as the first year in the graph). The scale indicates the sign and strength of the correlations.

Chapter 4 Summer temperature dependency of larch budmoth outbreaks

The highest correlations were achieved when the July to August temperatures were averaged prior to the calculation: correlationcoefficients between temperature and δ^{13}C are clearly higher (r = 0.68) than those between temperature and δ^{18}O (r = 0.49). Previous year conditions do not have a significant effect on either carbon or oxygen values, as there are no significant correlations between the isotope series and the months of the previous year (not shown in Fig. 4.1). In addition to finding high positive correlations with temperature, we also found negative correlations with precipitation (not shown in Fig. 4.1). This result was expected, as temperature and precipitation are anti-correlated climate parameters in this region.

To test the temporal stability of the climate-isotope relationship, we calculated 40-year running correlations between the monthly temperature records and both isotope series (Fig. 4.1b). In general, we found positive correlations with temperature for the entire century, which is consistent with the finding above. The temperature signal in both isotope series is mainly restricted to July and August of the current year. Although the correlation coefficients are slightly higher between δ^{13}C and temperature than between δ^{18}O and temperature, temporal stability is given. Nevertheless, no distinct temporal shifts in the climate-isotope relationship are indicated.

4.3.2 Tree-growth, LBW outbreaks, and climate

To determine the influence of LBM outbreaks on various tree-ring parameters without a climatic bias, we calculated the residuals between TRW, δ^{13}C and δ^{18}O of the CHRONO samples and the July to August temperature for the period AD 1900–2004 by plotting the tree-ring parameters against climate variables and fitting a regression line (Fig. 4.2). The LBM event years are highlighted on Fig. 4.2 to characterize their distribution within all sample years. While for TRW the LBM outbreaks are obviously outliers compared to all years, the δ^{13}C and δ^{18}O values of the LBM event years do not show a significantly different distribution compared to the entire data set. The P_{wilcox} value for δ^{13}C and δ^{18}O is higher by a factor of 10–12 compared to the threshold ($P_{wilcox} < 0.05$), but the TRW-temperature relationship is noticeably lower (compare P_{wilcox} values in Fig. 4.2), indicating a significant difference between the LBM event years and non-affected years for TRW, but not for both isotopic ratios.

4.3 Results

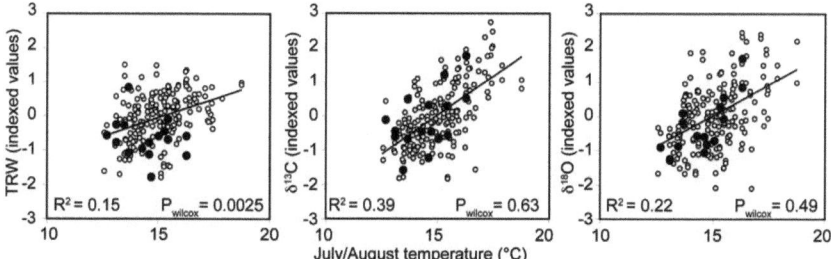

Figure 4.2: Analysis of residuals of tree-ring width (TRW, left), $\delta^{13}C$ (middle) and $\delta^{18}O$ (right) of the chronology [averaged over southfacing (S) and northfacing (N) sites] with July/August temperature for the period AD 1900–2004. All values were standardized to the same mean (=0) and standard deviation (= 1). Closed circles Gray larch budmoth (LBM: *Zeiraphera diniana* Gn.) events (n = 16). P_{wilcox} values of the Wilcoxon-Mann-Whitney rank sum test indicate significant differences between groups if P_{wilcox} <0.05.

4.3.3 Site-specific aspects of LBM dynamics (AD 1900–2004)

Results of the SEA (Fig. 4.3) showed a significant decrease of TRW during the outbreak year, which was even more distinct in the year thereafter for all stands (CHRONO, LOE, SIM). Despite this obvious effect on TRW, hardly any effect was evident from the isotope parameters. The SIM site, in particular, does not show any response of stable isotope signatures during LBM outbreaks. The $\delta^{13}C$ decrease during LBM outbreaks at the CHRONO and LOE sites is scarcely visible and within the error margins, and the minor effect in the case of $\delta^{18}O$ does not exceed the overall amplitude of the signal.

Although some outbreak events were detected on only one of the slopes, the response was identical to those that occurred on all slopes. Comparisons of the northern versus southern aspects of the stands revealed no difference in terms of LBM outbreaks for any of the three parameters. The larger variability within the LOE and SIM sites compared to the CHRONO site may be due to a more individualistic response of younger trees to LBM outbreaks (Esper et al., 2007).

Chapter 4 Summer temperature dependency of larch budmoth outbreaks

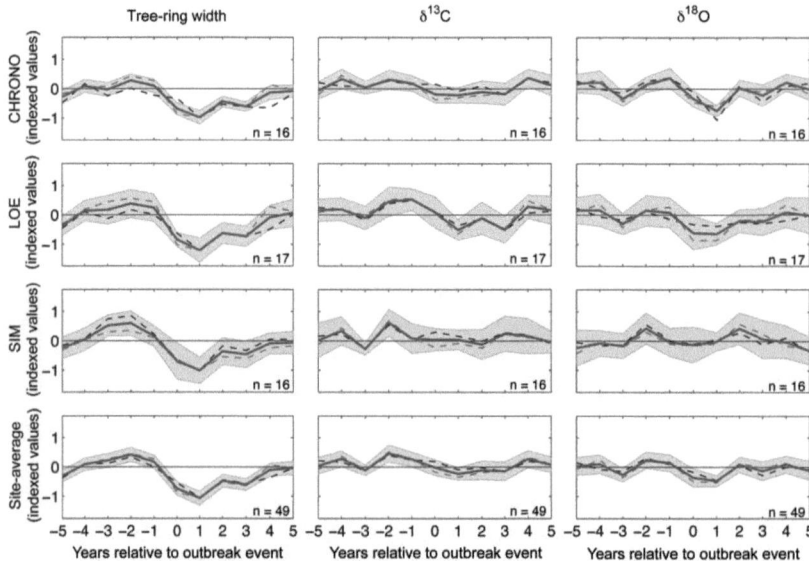

Figure 4.3: Outbreak patterns in tree-ring data (tree-ring width, δ^{13}C, δ^{18}O) of the three sites (CHRONO, long-term chronology sites in the Lötschental region; LOE, younger material from the Lötschental region; SIM, younger material from the Simplon region) and the average of all sites for AD 1900–2004. All values were standardized to the same mean (= 0) and standard deviation (= 1). They are centered on the outbreak event (year 0) and show values up to 5 years before and after these events. The red curve is an average of all outbreaks within one stand (n = number of events averaged), combining data on the southern slopes (green curve) and northern slopes (blue curve). Gray shadings indicate ± 2 SE of the averaged red curve. The horizontal black line denotes the average over the 11-year window.

4.3.4 Species-specific response

For the period AD 1946–1995, we compared isotopic data of LBM non-host spruce (*Picea abies*; compare Treydte et al., 2001) with our LBM host data from the Lötschental (CHRONO, LOE-S, LOE-N; Table 4.3). The spruce samples are about the same age as the LOE-S and LOE-N stands (150–250 years), whereas the CHRONO samples are older (450–550 years). Surprisingly, spruce correlated with larch for $\delta^{13}C$ in the same range ($r_{CHRONOLOGY} = 0.82$, $r_{LOE-S} = 0.72$, $r_{LOE-N} = 0.69$) as the larch sites were correlating among each other (between r = 0.87 and 0.69). This indicates that inter-species relationships were as strong as intra-species relationships at the same site. Oxygen isotopes provided slightly weaker correlation coefficients between spruce and larch than among the larch sites; nevertheless, the oxygen correlations between spruce and larch were still highly significant (P < 0.001), indicating a strong relationship between larch (host) and spruce (non-host). Beyond these findings, we evaluated the host/non-host behavior of the carbon and oxygen isotopes (Fig. 4.4) during LBM event years by plotting spruce and larch isotope time series with normalized ordinates to account for the natural inter-species isotopic offset. During LBM outbreak years (1954, 1963, 1972 and 1981), host and non-host data revealed the same trend for carbon as well as for oxygen isotopes, showing again that isotopic signatures were not affected by LBM.

Table 4.3: Correlation matrices between larch (CHRONO, LOE-S,LOE-N) and spruce for carbon isotopes and oxygen isotopes

		CHRONO	LOE-S	LOE-N
$\delta^{13}C$	LOE-S	0.87***		
	LOE-N	0.80***	0.69***	
	SPRUCE	0.82***	0.72***	0.69***
$\delta^{18}O$	LOE-S	0.89***		
	LOE-N	0.77***	0.73***	
	SPRUCE	0.68***	0.63***	0.54***

All correlation coefficients are highly significant (***P < 0.001)

Chapter 4 Summer temperature dependency of larch budmoth outbreaks

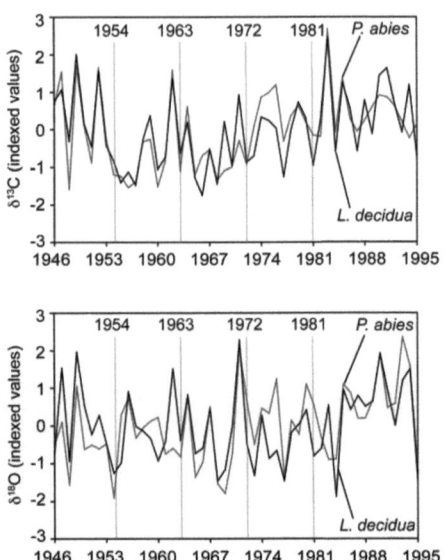

Figure 4.4: A comparison of larch (black line; *Larix decidua* Mill.) and spruce (red line; *Picea abies*) δ^{13}C (upper panel) and δ^{18}O (lower panel) data in the Lötschental for the period AD 1995–1946 (maximum length of spruce data). All isotope values were standardized to the same mean (= 0) and standard deviation (= 1). Vertical bars specify LBM event years (n = 4).

4.3.5 Long-term climate forcing and LBM outbreaks

Figure 4.5 shows all of the LBM event years (n = 26) of the period AD 1660–2000 for the TRW (Fig. 4.5b), δ^{13}C (Fig. 4.5c), and δ^{18}O (Fig. 4.5d) values of the chronology samples and compares these to the averaged July to August temperature reconstruction of the corresponding grid cell (Casty et al., 2005; Fig. 4.5a) of these outbreak years. As noted above for the 20th century, TRW in the outbreak years is significantly below average, with only one exception (1896), whereas this is not the case for both isotope ratios. Interestingly, we observed that the July to August temperature in LBM event years was significantly lower than the long-term average value (P_{wilcox} = 0.004), indicating that cool summers are conducive to LBM outbreaks. Such an anomaly in

4.3 Results

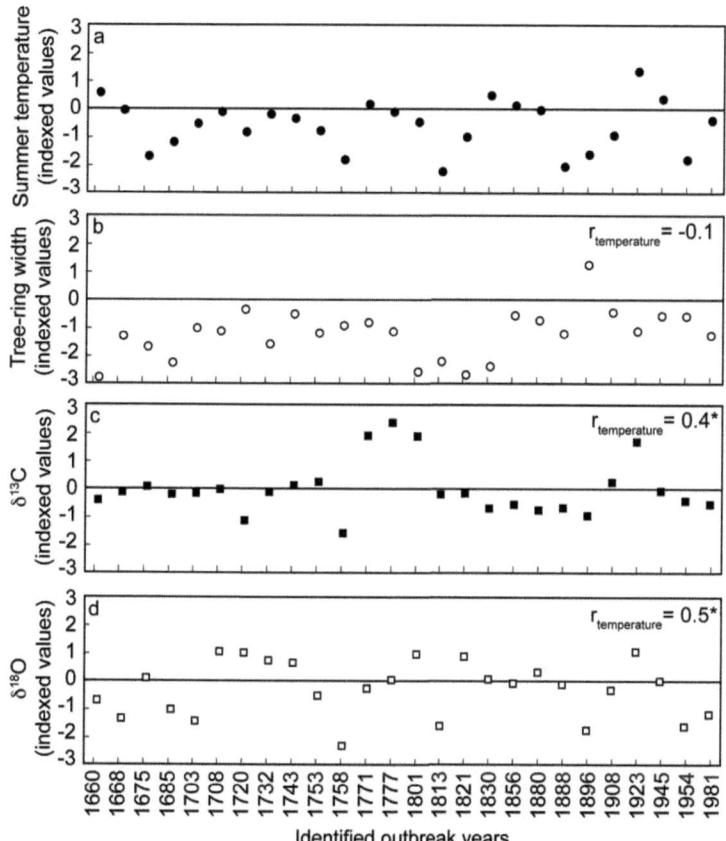

Figure 4.5: All 26 outbreak years and tree-ring parameters for AD 1660–2000: a) July to August temperature according to Casty et al. (2005), b) tree-ring width, c) $\delta^{13}C$, d) $\delta^{18}O$. All values were standardized to the same mean (=0) and standard deviation (=1). Correlation coefficients represent the relationship between tree-ring variables and reconstructed temperature (*P < 0.05). The horizontal black lines indicate the average over the 340-year period.

Chapter 4 Summer temperature dependency of larch budmoth outbreaks

weather conditions during LBM event years was found only for the July to August temperature, while the temperature data of the other months were not found to be much different from longterm averages (data not shown). Apparently, the isotope ratios still respond to deviations in summer temperature during LBM years; in particular, the cold summer of 1758 is evidently reproduced in the signature of both isotopes. The correlation between isotope values and July to August temperature for these outbreak years is significant (P < 0.05) for δ^{13}C as well as for δ^{18}O, similar to the results found for the calibration period for all years (see above). Interestingly, the correlation coefficient between TRW and temperature, when only LBM outbreak years are used, turned out to be slightly negative (not significant), whereas one would expect a positive correlation between the two variables. This shows that the relationship between TRW and temperature is disturbed during years of LBM defoliation whereas the isotope-temperature relationship remains distinct.

4.4 Discussion

We have provided several lines of evidence that LBM events do not affect the isotope ratios in tree rings of larch, neither during nor after an outbreak year:

- The climate-isotope relationship of the calibration period (AD 1900–2004) reveals the expected results despite several strong outbreaks during this period: correlations are restricted to the vegetation period with robust, consistently positive correlations for temperature (in agreement with theoretical expectations; see McCarroll and Loader, 2004). The magnitude of the correlation coefficients is amongst the highest yet reported. They are somewhat higher for carbon than for oxygen isotopes, with relationships being stable over the 20th century.

- When climate variations are removed by calculating residuals, there is no statistically significant difference in isotope values between years affected by LBM outbreaks and other years, whereas for TRW the differences are significant.

- SEA (reflecting an average response of all outbreaks) shows distinct decreases of TRW during LBM events (and the following years); the amplitudes of δ^{13}C and δ^{18}O do not provide such a distinct deviation. This result is confirmed by trees of two different age classes and several sites that cover two different aspects (N vs.

4.4 Discussion

S) and located up to 50 km apart. The similar behavior of trees on north- and south-facing slopes is to be expected, as exposure has little effect on temperatures that are relevant for trees at such elevations (Körner and Paulsen, 2004).

- Further evidence is provided by the high correlation coefficients between larch (host) and spruce (non-host) for both isotope series as well as by the similar curve patterns of the chronologies of different species from the same site in the absence of insect impacts. This similar behavior is not self-evident, because the species may differ in their sensitivity to climate (Saurer et al., 2008; Treydte et al., 2007). The strong climatic limitation of tree growth at our high-altitude study sites may have a similar limitation effect on the isotope variations and thus result in the strong correlation between larch and spruce, even during LBM defoliation events.

The negligible effect of LBM infestations on tree-ring isotope ratios in larch, as shown above, is rather surprising, as even subtle changes in leaf physiology, such as stomatal closure (e.g., during water shortage), are known to strongly affect both carbon and oxygen isotopes (Barbour et al., 2004; Leavitt and Long, 1988). The physiological response of conifers to insect-related defoliation could therefore significantly impact the rate of photosynthetic assimilation. Based on the hypothesis of an increased photosynthetic rate as a compensatory mechanism for defoliation, this would trigger a $\delta^{13}C$ enrichment in tree-ring cellulose. The synchronous behavior of carbon and oxygen isotopes during LBM defoliation events suggests that isotope composition depends directly on climatic conditions. This synchronicity is present not only within species at different sites but also between host (larch) and non-host (spruce) species, thereby confirming that regional climate variations are primarily responsible for the isotopic composition.

We assume that during the main defoliation event itself, hardly any photosynthesis takes place due to either a lack of needles or desiccated or dysfunctional (brown) needles. Accordingly, very little photosynthates are produced and, thus, hardly any tree-ring cellulose is built in which a climatic signal could be incorporated. Consequently, the tree-ring cellulose of LBM outbreak years is formed either before the heavy feeding occurs or shortly thereafter, i.e., when larch trees refoliate 3–4 weeks after defoliation (Baltensweiler et al., 2008). The refoliation could explain the apparent phase shift between the main production of wood during outbreak years and the observed high correlations with July to August climate conditions, although it is difficult to establish a causal link. This theory is in agreement with results from previous studies on the impact of phytophagous insects on the carbon isotope composition of plant material showing a lack of infestation response on photosynthetic discrimination and carbon isotopic composition of the remaining leaf material and tree-ring cellulose (Ellsworth et al., 1994; Haavik et al., 2008).

Our unaffected isotopic signatures in tree rings during LBM outbreak events seem to be contrary to findings by Simard et al. (2008) who investigated another Lepidopteran species, the eastern spruce budworm (ESB, *Choristoneura fumiferana* Clem.). Their comparison of host species (*Abies balsamea* and *Picea mariana*) and non-host species (*Pinus banksiana*) revealed carbon isotope enrichments during two outbreak episodes, while oxygen isotopes remained unaffected. Although species of the same order, LBM and ESB have different life cycles: while LBM defoliation events are strongly coupled with one vegetation period and reoccur with a periodicity of 8- to 10-year intervals, ESB events appear in decadal scale frequency (32- to 34-year intervals) and last for 5 years and more. If trees are affected for several following years, they need to mobilize reserves to maintain functionality. Carbon allocation, however, can alter the isotopic signatures in tree rings in a characteristic manner (Keel et al., 2006). Thus, it may well be that ESB events influence the isotope signal while LBM outbreaks do not.

The absence of an isotope effect during LBM defoliation events is in contrast to the TRW response patterns. A narrow ring is formed not only during the outbreak year itself, but often-even more pronounced- in subsequent years (up to 6 years). Accordingly, the LBM effect dominates TRW relative to the climatic influence during these periods. Detecting a climate signal in TRW during LBM event years is still possible, but corrections are needed, such as a replacement of LBM-induced negative growth depressions with information derived from non-affected trees (Büntgen et al., 2006b; Esper et al., 2007).

When a long climate record covering more than three centuries is considered (Casty et al., 2005), cold summers are often related to severe LBM outbreaks, and the isotope ratios reliably mirror these relatively cool conditions. Based on TRW only, this assumption would lead to circular reasoning because LBM detection is predicated on the identification of narrow TRW, which may, alternatively, be caused by cold summers. For further validation, we additionally calculated July to August temperatures (Casty

4.4 Discussion

et al., 2005) of the LBM outbreak events determined by Baltensweiler and Rubli (1999) for the Valais and Engadin regions for the period AD 1850–1990 and compared these to average July to August temperatures of AD 1660–2000. Although the number of identified outbreak events is, compared to our dataset, lower (according to the terminology by Baltensweiler and Rubli (1999) medium and heavy discoloration events were considered), we demonstrate significantly lower (P < 0.05) July to August temperatures during LBM event years compared to the long-term average for the Engadin ($P_{wilcox} = 0.011$) and the Valais regions ($P_{wilcox} = 0.024$), thus confirming our discovery.

This highly significant finding over a period of 340 years indicates that cool July to August temperatures promote LBM outbreaks -even if the latter are occurring at highly regular intervals. This is rather surprising as outbreaks are usually triggered earlier in the vegetation period and are likely to be linked to subcontinental climate dependencies of LBM population cycles: one dominating factor for an optimal development of LBM population cycles is the duration of the winter diapause, which is highly conditioned by annual temperature profiles. Two aspects are crucial for the survival of the 2-mm-long first instar larvae: the fixed amount of energy for egg development and hatching provided at the time of oviposition (usually in August) and the hatching of the larvae 9–11 months later simultaneously with the flushing of the larch trees. Ideal requirements to fulfill these criteria are long, cold winters with more than 120 days below 2 °C. If spring and summer temperatures are high, the ontogenetic development from egg to moth is terminated soon, leading to an early diapause with fewer frost days, resulting in an elevated egg mortality (Baltensweiler, 1993b; Baltensweiler et al., 1977). This finding is consistent with our discovery that devastating LBM events are coupled with low late-summer temperatures.

Another potential explanation for the ecological significance of low summer temperatures is the link between needle quality and summer temperatures. One basic regulatory mechanism of the LBM cycle is the induced change in food quality for two or more subsequent larval generations (Baltensweiler, 1993b). Larch as a deciduous species re-grows its needles every year. Therefore, the LBM has to adapt its development to needle maturation and has to cope with large changes in food quality (e.g., raw fiber and protein content). Synchronization of larvae evolution and needle maturation is a prerequisite for outbreak events (Asshoff and Hattenschwiler, 2006; Baltensweiler et al., 1977). Slight asynchrony of these fine-tuned events may result in serious consequences for species interaction. If summer temperatures are a driving force for needle maturation, they indirectly have an impact on the LBM cycle as an LBM outbreak can occur only if needle maturation is in optimal agreement with LBM larvae development. Poor food quality, induced by above-average summer temperatures, leads to heavy larval and pupal mortality and reduced fertility, as observed by Baltensweiler et al. (1977) and Fischlin (1982). However, it has recently been suggested that LBM population cycles may be driven by parasitoids rather than changes in needle quality (Berryman, 1996; Turchin et al., 2003). Therefore, an interaction between late-summer temperatures and parasitoids with an indirect effect on LBM populations cannot be ruled out completely.

The reconstructed persistence and regularity of LBM reoccurrence over the past 1200 years is remarkable (Esper et al., 2007). However, since the early 1980s no alpine-wide synchronized LBM outbreak event has occurred. Locally, LBM populations have attained sub-defoliating peak densities only, suggesting a diminishment of the oscillation amplitude, but not necessarily an alteration of the period. This absence of massive outbreaks is the longest detected within the last 1200 years. Simultaneously, temperature reconstructions from the European Alps reveal that conditions during the late 20th century represent the warmest period of the past millennium (Büntgen et al., 2006b; IPCC, 2007). This coincidence together with our finding of July to August temperature dependencies of LBM outbreaks leads us to conclude that high summer temperatures can dramatically impair LBM dynamics. Similar observations have been made for oak-winter moth phenology (another Lepidopteran species), which has been significantly disturbed by recently higher spring temperatures (Visser and Holleman, 2001). It remains to be seen how this recent warming trend will influence the prominent forest disturbance phenomenon of LBM infestations and how it will alter ecosystem processes in subalpine larch forests. We suggest that stable isotopes in tree rings in combination with TRW are a powerful tool to study the impact of climate on insect infestations over centuries to millennia, provided that isotope ratios are not affected by the outbreaks but by climate only, as was shown for the LBM in this study.

In conclusion, we would like to emphasize that besides our discovery of late-summer dependencies of LBM cycles, the preservation of a climatic signal in the isotopic signature of tree rings during LBM outbreak events is striking. While larch TRW and density contain a strong climatic signal after appropriate corrections for LBM infestations, a more complete retrieval of information from this unique long-living archive is provided by the additional consideration of stable isotope ratios.

4.4 Discussion

Acknowledgements

This work was funded by the EU project FP6-2004-GLOBAL-017008-2 (MILLENNIUM). Thanks to A. Verstege and D. Nievergelt for support in the Dendro-LAB, to M. Tröndle and L. Läubli for assistance with sample preparation and to G. Helle for the oxygen measurements of the spruce samples. Many thanks also to U. Baltensperger and D. McCarroll for the valuable discussions and helpful comments. The experiments comply with the current laws in Switzerland.

References

Adams, J. B., Mann, M. E., and Ammann, C. M. (2003). Proxy evidence for an El Nino-like response to volcanic forcing. *Nature*, 426(6964):274–278; doi:10.1038/nature02101.

Anderson, R. M. and May, R. M. (1980). Infectious-diseases and population-cycles of forest insects. *Science*, 210(4470):658–661; doi:10.1126/science.210.4470.658.

Anderson, W. T., Bernasconi, S. M., McKenzie, J. A., and Saurer, M. (1998). Oxygen and carbon isotopic record of climatic variability in tree ring cellulose (*Picea abies*): an example from central Switzerland (1913-1995). *Journal of Geophysical Research*, 103(D24):31625–31636; doi:10.1029/1998JD200040.

Asshoff, R. and Hattenschwiler, S. (2006). Changes in needle quality and larch budmoth performance in response to CO_2 enrichment and defoliation of treeline larches. *Ecological Entomology*, 31(1):84–90; doi:10.1111/j.0307-6946.2006.00756.x.

Auer, I., Böhm, R., Jurkovic, A., Lipa, W., Orlik, A., Potzmann, R., Schoner, W., Ungersbock, M., Matulla, C., Briffa, K., Jones, P., Efthymiadis, D., Brunetti, M., Nanni, T., Maugeri, M., Mercalli, L., Mestre, O., Moisselin, J. M., Begert, M., Muller-Westermeier, G., Kveton, V., Bochnicek, O., Stastny, P., Lapin, M., Szalai, S., Szentimrey, T., Cegnar, T., Dolinar, M., Gajic-Capka, M., Zaninovic, K., Majstorovic, Z., and Nieplova, E. (2007). HISTALP - historical instrumental climatological surface time series of the Greater Alpine Region. *International Journal of Climatology*, 27(1):17–46; doi:10.1002/joe.1377.

Baltensweiler, W. (1993a). A contribution to the explanation of the larch bud moth cycle, the polymorphic fitness hypothesis. *Oecologia*, 93(2):251–255; doi:10.1007/BF00317678.

Baltensweiler, W. (1993b). Why the larch bud moth cycle collapsed in the subalpine larch-cembran pine forests in the year 1990 for the first time since 1850. *Oecologia*, 94(1):62–66; doi:10.1007/00317302.

Baltensweiler, W., Benz, G., Bovey, P., and Delucchi, V. (1977). Dynamics of larch bud moth populations. *Annual Review of Entomology*, 22:79–100; doi:10.1146/annurev.en.22.010177.

Baltensweiler, W. and Rubli, D. (1999). Dispersal: an important driving force of the cycling population dynamics of the larch budmoth, *Zeiraphera diniana* Gn. In Swiss Federal Institute for Forest, S. and Landscape Research (WSL), B., editors, *Forest snow and landscape research*, volume 74, page 153. Paul Haupt, Berne, Stuttgart, Vienna.

Baltensweiler, W., Weber, U. M., and Cherubini, P. (2008). Tracing the influence of larch-bud-moth insect outbreaks and weather conditions on larch tree-ring growth in Engadine (Switzerland). *Oikos*, 117(2):161–172; doi:10.1111/j.2007.0030-1299.16117.x,.

Barbour, M. M., Roden, J. S., Farquhar, G. D., and Ehleringer, J. R. (2004). Expressing leaf water and cellulose oxygen isotope ratios as enrichment above source water reveals evidence of a Péclet effect. *Oecologia*, 138(3):426–435; doi:10.1007/s00442-003-1449-3.

Berryman, A. A. (1996). What causes population cycles of forest Lepidoptera? *Trends in Ecology and Evolution*, 11(1):28–32; doi:10.1016/0169-5347(96)81066-4.

Biondi, F. and Waikul, K. (2004). Dendroclim2002: A C++ program for statistical calibration of climate signals in tree-ring chronologies. *Computers and Geosciences*, 30(3):303–311; doi:10.1016/j.cageo.2003.11.004.

Boettger, T., Haupt, M., Knoller, K., Weise, S. M., Waterhouse, J. S., Rinne, K. T., Loader, N. J., Sonninen, E., Jungner, H., Masson-Delmotte, V., Stievenard, M., Guillemin, M. T., Pierre, M., Pazdur, A., Leuenberger, M., Filot, M., Saurer, M., Reynolds, C. E., Helle, G., and Schleser, G. H. (2007). Wood cellulose preparation

4.4 Discussion

methods and mass spectrometric analyses of $\delta^{13}C$, $\delta^{18}O$ and nonexchangeable $\delta^{2}H$ values in cellulose, sugar, and starch: an interlaboratory comparison. *Analytical Chemistry*, 79(12):4603–4612; doi:10.1021/ac0700023.

Büntgen, U., Esper, J., Frank, D. C., Nicolussi, K., and Schmidhalter, M. (2005). A 1052-year tree-ring proxy for Alpine summer temperatures. *Climate Dynamics*, 25(2-3):141–153; doi:10.1007/s00382-005-0028-1.

Büntgen, U., Frank, D. C., Bellwald, I., Kalbermatten, H., Freund, H., Schmidhalter, M., Bellwald, W., Neuwirth, B., and Esper, J. (2006a). 700 years of settlement and building history in the Lötschental/Valais. *Erdkunde*, 60:96–112; doi:10.3112/erdkunde.2006.02.02.

Büntgen, U., Frank, D. C., Niervergelt, D., and Esper, J. (2006b). Summer temperature variations in the European Alps, AD 755-2004. *Journal of Climate*, 19(21):5606–5623; doi:10.1175/JCLI3917.1.

Casty, C., Wanner, H., Luterbacher, J., Esper, J., and Böhm, R. (2005). Temperature and precipitation variability in the European Alps since 1500. *International Journal of Climatology*, 25:1855–1880; doi:10.1002/joc.1216.

Ellsworth, D. S., Tyree, M. T., Parker, B. L., and Skinner, M. (1994). Photosynthesis and water-use efficiency of sugar maple (*Acer saccharum*) in relation to pear thrips defoliation. *Tree Physiology*, 14(6):619–632; doi:10.1093/treephys/14.6.619.

Esper, J., Büntgen, U., Frank, D. C., Niervergelt, D., and Liebhold, A. (2007). 1200 years of regular outbreaks in alpine insects. *Proceedings of the Royal Society B*, 274:671–679; doi:10.1098/rspb.2006.0191.

Esper, J., Cook, E. R., and Schweingruber, F. H. (2002). Low-frequency signals in long tree-ring chronologies for reconstructing past temperature variability. *Science*, 295(5563):2250–2253; doi10.1126/science.1066208.

Farquhar, G. D., O'Leary, M. H., and Berry, J. A. (1982). On the relationship between carbon isotope discrimination and the intercellular carbon dioxide concentration in leaves. *Australian Journal of Plant Physiology*, 9(2):121–137.

Fischlin, A. (1982). *Analyse eines Wald-Insekten-Systems: Der Subalpine Lärchen-Arvenwald un der Graue Lärchenwickler Zeiraphera diniana Gn. (Lepidoptera, Tortricidae)*. Dissertation number 6977, ETH, Zürich.

Frank, D. and Esper, J. (2005). Characterisazion and climate response patterns of a high-elevation multi-species tree-ring network in the European Alps. *Dendrochronologia*, 22:107–121; doi:10.1016/j.dendro.2005.02.004.

Guiot, J. (1991). The bootstrapped response function. *Tree-Ring Bulletin*, 51:39–41.

Haavik, L. J., Stephen, F. M., Fierke, M. K., Salisbury, V. B., Leavitt, S. W., and Billings, S. A. (2008). Dendrochronological parameters of northern red oak (*Quercus rubra* L. (Fagaceae)) infested with red oak borer (*Enaphalodes rufulus* (Haldeman) (Coleoptera : Cerambycidae)). *Forest Ecology and Management*, 255(5-6):1501–1509; doi:10.1016/j.foreco.2007.11.005.

Holmes, R. L. (1983). Computer-assisted quality control in tree-ring dating and measurements. *Tree-Ring Bulletin*, 43:69–78.

IPCC (2007). *Climate Change 2007: the Physical Science Basis. Contribution of Working Group I to the Fourth Assessment Report of the Intergovernmental Panel on Climate Change*. Cambridge University Press, Cambridge, United Kingdom and New York, NY, USA.

Keel, S. G., Siegwolf, R. T. W., and Körner, C. (2006). Canopy CO_2 enrichment permits tracing the fate of recently assimilated carbon in a mature deciduous forest. *New Phytologist*, 172(2):319–329; doi:10.1111/j.1469-8137.2006.01831.x.

Körner, C. and Paulsen, J. (2004). A world-wide study of high altitude treeline temperatures. *Journal of Biogeography*, 31:713–732; doi:10.1111/j.1365-2699.2003.01043.x.

Kurz, W. A., Dymond, C. C., Stinson, G., Rampley, G. J., Neilson, E. T., Carroll, A. L., Ebata, T., and Safranyik, L. (2008). Mountain pine beetle and forest carbon feedback to climate change. *Nature*, 452(7190):987–990; doi:10.1038/nature06777.

Leavitt, S. W. and Long, A. (1986). Influence of site disturbance on $\delta^{13}C$ isotopic time series from tree rings. In Jacoby, G. and Hornbeck, J., editors, *Proceedings of the International Symposium on Ecological Aspects of Tree-Ring Analysis, August 17-21, 1986*. Marymount College, Tarrytown, New York.

Leavitt, S. W. and Long, A. (1988). Stable carbon isotope chronologies from trees in the southwestern United States. *Global Biogeochemical Cycles*, 2(3):189–198; doi:10.1029/GB002i003p00189.

4.4 Discussion

Leuenberger, M. (2007). To what extent can ice core data contribute to the understanding of plant ecological developments of the past? In Dawson, T. and Siegwolf, R., editors, *Stable Isotopes as Indicators of Ecological Change*, pages 211–233. Elsevier Academic Press, London.

Mattson, W. J. and Addy, N. D. (1975). Phytophagous insects as regulators of forest primary production. *Science*, 190(4214):515–522.

McCarroll, D. and Loader, N. J. (2004). Stable isotopes in tree rings. *Quaternary Science Reviews*, 23(7-8):771–801; doi:10.1016/j.quascirev.2003.06.017.

Mitchell, T. D. and Jones, P. D. (2005). An improved method of constructing a database of monthly climate observations and associated high-resolution grids. *International Journal of Climatology*, 25(6):693–712; doi:10.1002/joc.1181.

Reynolds-Henne, C. E., Siegwolf, R. T. W., Treydte, K. S., Esper, J., Henne, S., and Saurer, M. (2007). Temporal stability of climate-isotope relationships in tree rings of oak and pine (Ticino, Switzerland). *Global Biogeochemical Cycles*, 21(4):GB4009; doi:10.1029/2007GB002945.

Roden, J. S., Lin, G., and Ehleringer, J. R. (2000). A mechanistic model for interpretation of hydrogen and oxygen isotope ratios in tree-ring cellulose. *Geochimica et Cosmochimica Acta*, 64(1):21–35; doi:10.1016/S0016-7037(99)00195-7.

Rolland, C., Baltensweiler, W., and Petitcolas, V. (2001). The potential for using *Larix decidua* ring widths in reconstructions of larch budmoth (*Zeiraphera diniana*) outbreak history: dendrochronological estimates compared with insect surveys. *Trees-Struct. Funct.*, 15(7):414–424; doi:10.1007/s004680100116.

Saurer, M., Cherubini, P., Reynolds-Henne, C. E., Treydte, K. S., Anderson, W. T., and Siegwolf, R. T. W. (2008). An investigation of the common signal in tree ring stable isotope chronologies at temperate sites. *Journal of Geophysical Research*, 113:G04035; doi:10.1029/2008JG000689.

Saurer, M. and Siegwolf, R. (2004). Pyrolysis techniques for oxygen isotope analysis of cellulose. In *Handbook of Stable Isotope Analytical Techniques*, volume 1, pages 497–508; doi:10.1016/B978-044451114-0/50025-9. Elsevier, New York.

Schweingruber, F. H. (1985). Dendroecological zones in the coniferous forests of Europe. *Dendrochronologia*, 3:67–75.

Chapter 4 Summer temperature dependency of larch budmoth outbreaks

Schweingruber, F. H., Eckstein, D., Serre-Bachet, F., and Bräker, O. U. (1990). Identification, presentation and interpretation of event years and pointer years in dendrochronology. *Dendrochronologia*, 8:9–38.

Simard, S., Elhani, S., Morin, H., Krause, C., and Cherubini, P. (2008). Carbon and oxygen stable isotopes from tree-rings to identify spruce budworm outbreaks in the boreal forest of Québec. *Chemical Geology*, 252(1-2):80–87; doi:10.1016/j.chemgeo.2008.01.018.

Stahel, W. (2002). *Statistische Datenanalyse: Eine Einführung für Naturwissenschaftler*. Vieweg, Wiesbaden, 4th edition.

Stokes, M. A. and Smiley, T. L. (1968). *An introduction to tree-ring dating.* (reprinted 1996). University of Arizona Press, Tucson, US, Chicago.

Treydte, K., Frank, D., Esper, J., Andreu, L., Bednarz, Z., Berninger, F., Boettger, T., D'Alessandro, C. M., Etien, N., Filot, M., Grabner, M., Guillemin, M. T., Gutierrez, E., Haupt, M., Helle, G., Hilasvuori, E., Jungner, H., Kalela-Brundin, M., Krapiec, M., Leuenberger, M., Loader, N. J., Masson-Delmotte, V., Pazdur, A., Pawelczyk, S., Pierre, M., Planells, O., Pukiene, R., Reynolds-Henne, C. E., Rinne, K. T., Saracino, A., Saurer, M., Sonninen, E., Stievenard, M., Switsur, V. R., Szczepanek, M., Szychowska-Krapiec, E., Todaro, L., Waterhouse, J. S., Weigl, M., and Schleser, G. H. (2007). Signal strength and climate calibration of a European tree-ring isotope network. *Geophysical Research Letters*, 34(24):L24302; doi:10.1029/2007GL031106.

Treydte, K. S., Schleser, G. H., Schweingruber, F. H., and Winiger, M. (2001). The climatic significance of $\delta^{13}C$ in subalpine spruces (Lötschental, Swiss Alps). *Tellus B*, 53(5):593–611; doi:10.1034/j.1600-0889.2001.530505.x.

Turchin, P., Wood, S. N., Ellner, S. P., Kendall, B. E., Murdoch, W. W., Fischlin, A., Casas, J., McCauley, E., and Briggs, C. J. (2003). Dynamical effects of plant quality and parasitism on population cycles of larch budmoth. *Ecology*, 84(5):1207–1214; doi:10.1890/0012-9658(2003)084[1207:DEOPQA]2.0.CO;2.

Visser, M. E. and Holleman, L. J. M. (2001). Warmer springs disrupt the synchrony of oak and winter moth phenology. *Proceedings of the Royal Society of London Series B-Biological Sciences*, 268(1464):289–294; doi:10.1098/rspb.2000.1363.

Weber, U. M. (1997). Dendroecological reconstruction and interpretation of larch budmoth (*Zeiraphera diniana*) outbreaks in two central Alpine valleys of Switzerland from 1470-1990. *Trees-Struct. Funct.*, 11(5):277–290; doi:10.1007/PL00009674.

Weidner, K., Helle, G., Löffler, J., Neuwirth, B., and Schleser, G. H. (2006). Stable isotope and tree-ring width variations of larch affected by larch budmoth outbreaks. In Haneca, K., Verheyden, A., Beeckman, H., Gärtner, H., Helle, G., and Schleser, G. H., editors, *TRACE - Tree Rings in Archaeology, Climatology and Ecology*, volume 5, pages 148–153. FZ Jülich, Jülich.

Yakir, D., DeNiro, M. J., and Ephrath, J. E. (1990). Effects of water-stress on oxygen, hydrogen and carbon isotope ratios in two species of cotton plants. *Plant, Cell and Environment*, 13(9):949–955; doi:10.1111/j.1365-3040.1990.tb01985.x.

5

A 350-year drought reconstruction from Alpine tree-ring stable isotopes

Anne Kress[1], Matthias Saurer[1], Rolf T.W. Siegwolf[1], David C. Frank[2], Jan Esper[3], and Harald Bugmann[4]

[1] *Laboratory of Atmospheric Chemistry, Paul Scherrer Institut, 5232 Villigen PSI, Switzerland*
[2] *Swiss Federal Research Institute WSL, 8903 Birmensdorf, Switzerland*
[3] *Department of Geography, Johannes Gutenberg University Mainz, 55099 Mainz, Germany*
[4] *Forest Ecology, Department of Environmental Sciences, Swiss Federal Institute of Technology Zürich (ETH), 8092 Zürich, Switzerland*

Global Biogeochemical Cycles 24:GB2011; doi:10.1029/2009GB003613, 2010

Chapter 5 A 350-year drought reconstruction from Alpine tree-ring stable isotopes

Abstract Climate reconstructions based on stable isotopes in tree rings rely on the assumption that fractionation-controlling processes are strongly linked to meteorological variables. In this context, we investigated the climate sensitivity of 350 years of carbon and oxygen isotope ratios of tree-ring cellulose from European larch obtained at a high-elevation site in the Swiss Alps (~2100 m a.s.l.). Unlike tree-ring width and maximum latewood density, which contain only summer temperature information at this site, we found that our stable isotope series reveal additionally to temperature a striking sensitivity to precipitation (mainly for carbon) and sunshine duration (mainly for oxygen) during July and August. A drought index reflecting the combined temperature and precipitation influence provided the most stable correlations over time for the carbon isotope series. All of these climate-isotope relationships are preserved in the isotope series obtained from younger trees at the same site, while strong inter-tree correlations further emphasize the high climate sensitivity. We thus present the first carbon isotope based summer drought reconstruction for the Swiss Alps, which provides new evidence for inter-annual to long-term changes in summer regional moisture variability from 1650 to 2004 in Europe, revealing extreme drought summers in the second half of the 18th century and throughout the 20th century.

Keywords

Carbon and oxygen isotopes – European larch – dendroclimatology

5.1 Introduction

For Europe, a dense network of suitable proxies for climate reconstruction exists with long instrumental records, detailed documentary evidence and highly temporally resolved natural archives. Despite the variety of proxy data, most climate reconstructions are focused on temperature, even though precipitation arguably plays a key role for human sustenance and economies as well as for many terrestrial ecosystems. Despite its importance, relatively little is known about past changes in precipitation regimes (e.g., Seager et al., 2007). Knowledge concerning the range of variability in hydroclimatic variables, such as regional precipitation, and the occurrence of both extreme events (droughts and floods) and long-term trends (multi-decadal to centennial-scale pluvial or drought periods), are of particular interest (e.g., Cook et al., 2004; Esper et al., 2007b).

5.1 Introduction

Continuous European hydroclimatic reconstructions covering the last few centuries are sparsely distributed and concentrated in southwestern Europe and the Mediterranean region (e.g., Esper et al., 2007b; Touchan et al., 2008). Long-term hydroclimatic records are particularly rare in central and northern Europe. Recent efforts to reconstruct precipitation and drought fluctuations for central Europe are based on documentary evidence (Brazdil et al., 2005), tree rings (Oberhuber and Kofler, 2002; Wilson et al., 2005; Büntgen et al., 2009) or multi-proxy approaches (Casty et al., 2005; Pauling et al., 2006). However, highly temporally resolved, documentary-based reconstructions may fail to capture low frequency signals, as descriptive evidence is discontinuous and biased by the perception of the observer (Brazdil et al., 2005), while the preservation of longer-term trends in tree-ring based climate reconstructions is potentially restricted by the need to remove the biological age-related trend (Cook et al., 1995) and may also be limited by an age-specific climate response (Esper et al., 2008). The diverse nature and characteristics of proxy records, their patchy spatial distribution over Europe, and high regional precipitation variability result in a challenging larger-scale comparison and interpretation of these records (Raible et al., 2006).

The existence of rather few tree-ring based hydroclimatic records in Europe reflects the limited capability of the common proxies to reliably capture a moisture signal. Climate reconstructions from tree-ring width and maximum latewood density, for example, are based upon annual cambial activity. At extreme sites (e.g., treeline sites), tree growth is usually limited by one climatic factor (e.g., temperature), while in temperate forests, the delineation of growth responses to a single controlling factor often fails (Friedrichs et al., 2009). Reliable hydroclimatic reconstructions from tree-ring evidence are thus restricted to very few moisture-sensitive sites (e.g., Wilson et al., 2005; Büntgen et al., 2009).

Stable carbon ($\delta^{13}C$) and oxygen ($\delta^{18}O$) isotopes in tree rings differ from the classical dendrochronological variables as they reflect physical conditions and tree responses rather than measuring net tree growth. While $\delta^{13}C$ values depend on variables affecting the photosynthetic uptake of CO_2 and are predominantly modulated by stomatal conductance and the rate of carboxylation during photosynthesis (Farquhar et al., 1989), $\delta^{18}O$ values are constrained by the isotopic ratio of the source water (Roden et al., 2000) and locally integrate stomatal conductance, which is coupled with transpiration (Yakir et al., 1990; Barbour et al., 2004). As these external and internal factors controlling isotopic signatures are fortunately closely correlated with meteorological variables,

Chapter 5 A 350-year drought reconstruction from Alpine tree-ring stable isotopes

such as e.g., relative humidity (Loader et al., 2008), stable isotopes may provide complementary climatic information in areas where the climate-ringwidth relationship is weak, or they may help to confirm long-term climatic trends inferred from tree-ring width or maximum latewood density series.

Numerous studies have demonstrated the potential of stable isotopes for the reconstruction of various climate variables, such as temperature (e.g., Gagen et al., 2007; Hilasvuori et al., 2009, precipitation (e.g., Treydte et al., 2006) and cloudiness (Young et al., 2010). Although climate reconstructions have been derived from well-replicated (e.g., Gagen et al., 2007) and millennial-length (e.g., Treydte et al., 2006) isotopic data, the attribution of a single climate variable to stable isotope series is complex and seems to be highly site dependent. Mostly a mixed climatic signal, typically temperature in combination with precipitation, appears to be reflected in isotope ratios (e.g., McCarroll and Pawellek, 2001; Treydte et al., 2001). Because several factors control isotope variability under most conditions, reconstructions of a single climate variable are prone to oversimplification (McCarroll and Loader, 2004), and palaeoclimate reconstructions of single climate variables from stable isotopes may therefore be of varying reliability (Reynolds-Henne et al., 2007). Nevertheless, recent isotope-network studies have demonstrated an astonishingly strong inter-site coherence for $\delta^{18}O$ (Treydte et al., 2007; Saurer et al., 2008), which may be related to atmospheric circulation patterns as the primary driving factor. Disentangling the role of the main controls of isotope fractionation and their associated meteorological variables is highly desirable in order to obtain reliable climate reconstructions from stable isotopes.

In this study, we assess the climate-isotope relationship for a 350-year carbon and oxygen isotope series obtained from tree-ring cellulose of European larch (*Larix decidua* Mill.) grown at tree-line sites in the Lötschental, Swiss Alps. Using these new isotope chronologies, we carry out a detailed sensitivity study of carbon and oxygen isotopes and regional climate. Based upon these analyses, we (1) provide new insights regarding the climate signal contained in carbon and oxygen series, and (2) present the first carbon-isotope based drought reconstruction for the European Alps. This reconstruction provides new evidence for inter-annual to long-term changes in regional summer-moisture variability from AD 1650 to 2004.

5.2 Methods

5.2.1 Study site and sampling strategy

The central alpine valley Lötschental is the largest side valley on the northern side of the Rhône river in the Valais, Switzerland (Fig. 5.1, Tab. 5.1). Its climate is characterized by the influence of the oceanic temperate-moist regime of the northern Alps and the dry subcontinental climate of the inner-alpine Rhône valley, which is one of the driest regions in Switzerland. Therefore, the climate can vary with high interannual differences in total precipitation, which may lead to sporadic strong droughts during the vegetation period (Treydte et al., 2001). The vegetation of the Lötschental is dominated by the subalpine belt of spruce-larch forests, which is gradually mixed with larch-Swiss Stone pine forests towards upper timberline, which is presently located between 2100 and 2200 m a.s.l.. The differences in treeline elevation in the current landscape are related to the varying intensity of alp-pasturing rather than climatic influences (Paulsen and Korner, 2001).

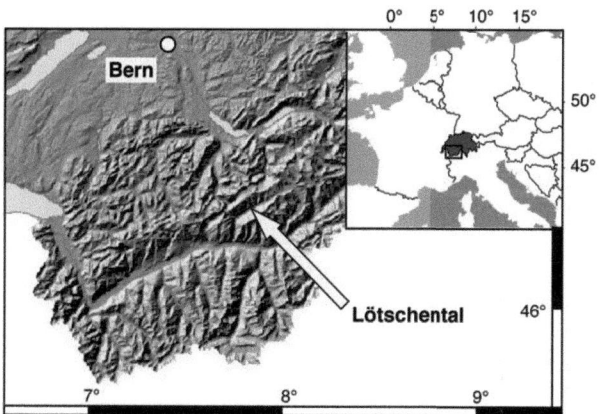

Figure 5.1: *Larix decidua* Mill. sampling site Lötschental situated in the Swiss Alps, Valais, Switzerland. Reproduced by permission of the Swiss Federal Office of Topography (JA082268).

Chapter 5 A 350-year drought reconstruction from Alpine tree-ring stable isotopes

The old-aged living larch and spruce trees, together with wood preserved in historical buildings has made the Lötschental an area of prime dendrochronological interest. Temperature reconstructions spanning the past millennium (Büntgen et al., 2005, 2006), dendroecological network analyses (Frank and Esper, 2005), isotopic site comparisons (Treydte et al., 2001), reconstructions of larch budmoth activity(Esper et al., 2007a; Kress et al., 2009a) and more recently the assessment of intra-annual tree growth along elevational transects (Moser et al., 2010) and intra-annual isotopic discrimination among the atmosphere-soil-plant systems (Boda et al., subm) have been performed.

In this study, samples of European larch (*Larix decidua* Mill.) were collected at two tree-line locations situated within the larch-Swiss Stone pine forests: at the SSE-facing slope (2100 m a.s.l.) above the village Blatten (1540 m a.s.l.) and the NNW-facing slope (2200 m a.s.l.) above the village Kippel (1376 m a.s.l.). The steep slopes are characterized by rather shallow soils. While at the SSE-facing, sunny slope podzolic cambisols are dominant, the NNW-facing, shady slope provides more humid soil conditions, i.e. ferric podzols (Treydte et al., 2001). At each site, trees were cored in two different age classes (450–550 years and 150–250 years), thereby care was taken to select trees under similar growth conditions (diameter at breast height, competition and microsite conditions). Each tree was cored twice at breast height (1,30 m) using a 5 mm increment borer.

Table 5.1: Site description of the Lötschental.

Site	Latitude (°N)	Longitude (°E)	Aspect	Slope	Altitude (m a.s.l.)	Soils[1]
Lötschental, S-facing slope	46°26'	7°48'	SSE	35°	2100	ferric podzols
Lötschental, N-facing slope	46°23'	7°47'	NNW	40°	2200	podzolic cambisols

[1] According to FAO soil classification

5.2.2 Sample analysis

Tree-ring widths were measured with a resolution of 0.01 mm using a semi-automated RinnTech system (Heidelberg, Germany) coupled to the TSAP tree-ring program, cross-dated (Stokes and Smiley, 1968) against a local master chronology (Büntgen et al., 2005) and verified by using the program COFECHA (Holmes, 1983). For isotope analysis five trees (two cores each) were chosen within the age class of 450-550 years, three of them from the SSE-facing slope and two from the NNW-facing slope respectively and eight trees within the younger age class (four from each slope), a number shown to be satisfactory to establish a representative isotope site record (e.g., McCarroll and Loader, 2004; Treydte et al., 2001).

Tree rings were split year-by-year for AD 1650–2004 (age-class 450-550 years) and AD 1900–2006 (age-class 150–250 years). Isotope analysis of 50 sub-samples of earlywood and latewood showed a very strong and coherent correlation between the two components, thus justifying whole-ring analysis (Kress et al., 2009b). This was fortunate as the very narrow rings would make ring-separation technically difficult and often leave insufficient material for isotopic measurements. As pooling rings of the same year prior to analysis was shown to retain climatic signals with little if any degradation and results in a great decrease in workload and necessary measurements (Treydte et al., 2007), we adopted two pooling approaches. The old trees were analyzed on a single-tree basis for AD 1901–2004 and using the so-called "10-year split pool approach" for AD 1650–1900, where years were pooled prior to analysis with single-tree measurements every 10th year. The break-up of the pool in regular intervals allows testing of the signal strength within the different trees. The younger trees were annually pooled prior to analysis for the entire period analyzed (AD 1900–2006).

Alpha-cellulose was extracted following standard procedures (Boettger et al., 2007) adapted for larch samples, homogenized by sonification, and freeze-dried for 24 h (Kress et al., 2009a). Carbon isotopic ratios were determined after combustion of samples to CO_2 at 1025°C in an elemental analyser (EA-1110; Carlo Erba Thermoquest, Milan, Italy) coupled to an isotope ratio mass spectrometer (Delta S or Delta Plus XL; Thermo Finnigan Mat, Bremen, Germany) with a reproducibility of 0.1‰. Oxygen isotopic ratios were analyzed with a reproducibility of 0.3‰ after pyrolysis to CO at 1080°C using a continuous flow method and a Delta Plus XP connected via a variable open-split interface Conflo III (both Thermo Finnigan Mat, Bremen, Germany) (Saurer and

Siegwolf, 2004). All isotope values are expressed in the delta notation relative to an international standard (VPDB for δ^{13}C and VSMOW for δ^{18}O). All carbon records were corrected for the atmospheric decline in δ^{13}C due to fossil fuel burning since the beginning of the industrialization (Leuenberger, 2007).

The signal strength of the carbon and oxygen isotope chronologies was assessed for the chronologies obtained within the 450–550 year age class by using the average inter-series correlation (RBAR) and the expressed population signal (EPS) statistics (Wigley et al., 1984). While RBAR is a measure of common variance between single series, independent of the number of series, EPS measures the degree to which the chronology approaches the theoretical population chronology from which it was drawn. Based upon the example from Wigley et al. (1984), the threshold for sufficient signal strength is often set to EPS > 0.85. Additionally we calculated the coefficient of variation (CV), a normalized measure of dispersion, to address the range of values within the chronologies.

5.2.3 Climate data

Climate calibration was performed using the HISTALP database, an alpine-wide meteorological dataset (Auer et al., 2007). We compared all four-isotope series with the existing temperature (n = 19), precipitation (n = 32) and sunshine duration (n = 11) series obtained from the Swiss part of the HISTALP database. For correlation analysis a mean of all climate series was taken for each climate variable, as various tests showed that correlations with the mean were equally strong or stronger than with any of the single climate series (see also Blasing et al., 1981). Correlations were calculated on a monthly basis from previous year March to current year December as well as seasonal averages for AD 1901–2004, using bootstrapped correlation analysis for significance testing (p < 0.001) (Guiot, 1991). By calculating running correlations between the monthly records and both isotope series over 40-year time windows, we tested the temporal stability of the climate-isotope relationship. In addition, spectral analyses were performed using the multi-taper method (MTM, 2 year resolution, 3 tapers) of the Spectra software (Ghil et al., 2002).

To further investigate spatial climate correlations, we used temperature and precipitation data of an updated version of the 0.5° x 0.5° monthly gridded meteorological dataset CRU TS 3 (Mitchell and Jones, 2005) as well as various monthly resolved self-calibrating Palmer Drought Severity Index (scPDSI) datasets available for the European Alps (10-minute latitudinal-longitudinal resolution) (van der Schrier et al., 2007), central Europe (0.5° x 0.5° grid, van der Schrier et al., 2006), and the globe (2.5° x 2.5° grid, Dai et al., 2004), accessed via the KNMI climate explorer (http://climexp.knmi.nl).

In addition, we assessed the relationships between different climate variables and the climate-isotope relationship prior to the 20th century by using the monthly resolved temperature and precipitation reconstruction of Casty et al. (2005) from the grid cell 46.25°N 7.75°E covering the Lötschental (AD 1660-2000), as well as a drought index derived from these reconstructions (see below).

5.2.4 Drought index

For AD 1901–2004 a simple drought index (DRI, Bigler et al., 2006) was calculated based on the temperature and precipitation amount series of the Swiss part of the HISTALP dataset (Auer et al., 2007):

$$DRI = P - PET \tag{5.1}$$

with P equal to the monthly precipitation amount and PET equal to the monthly sum of estimated potential evapotranspiration as a function of monthly mean temperatures and geographical latitude, following the formulation of Thornthwaite (1948). DRI values were calculated on a monthly basis. We used the simple DRI rather than a more mechanistic drought index for several reasons: (1) Bigler et al. (2006) showed no significant improvement compared to the more complex drought index by Bugmann and Cramer (1998); (2) no assumptions about detailed parameters such as soil water holding capacity are necessary for calculating DRI, and (3) a drought index that captures wet and dry extremes rather than just indicating dry extremes was more suitable to fully characterize moisture variability. DRI values are easily interpretable as regional water availabilities; values below zero indicate moisture deficits whereas values above zero indicate water excess.

DRI signal analyses were conducted for high- and low-pass filtered data using cubic

Chapter 5 A 350-year drought reconstruction from Alpine tree-ring stable isotopes

smoothing splines with a 50% frequency response cut-off at 10 years (Cook and Peters, 1981). Based on monthly correlation results, a July-August DRI reconstruction was built from δ^{13}C using simple linear regression. To assess the predictive skill, explained variance (R^2) of the regression models, reduction of error statistic (RE) (Fritts, 1976), coefficient of efficiency (CE) (Cook et al., 1994) and the Durbin-Watson statistic (DW) (Durbin, 1951) were systematically compared for independent calibration and verification periods. While RE and CE are measures of shared variance between target and proxy series and generally lower than R^2 (Fritts, 1976; Cook et al., 1994), the DW tests for first-order autocorrelations in the model residuals. A DW-value of 2 indicates no first-order autocorrelation in the residuals, whereas a value of DW > 2 (DW < 2) suggests negative (positive) autocorrelation (Durbin, 1951). Reconstruction errors were estimated by the root mean squared error (RMSE) of the prediction (Wilks, 2006) divided by the 51-year running RBAR (maximum value within each 51-year window) of the carbon isotope series.

Lastly, comparisons were performed with three different reconstructions of spring/summer precipitation from central Europe (Oberhuber and Kofler, 2002; Wilson et al., 2005; Wimmer in Büntgen et al., 2009) and one recent summer drought reconstruction from Slovakia (Büntgen et al., 2009). All these reconstructions are annually resolved and based on tree-ring widths.

5.3 Results

5.3.1 Signal strength in carbon and oxygen isotope series

Annually resolved δ^{13}C and δ^{18}O chronologies obtained from older (450–550 years) and younger (150–250 years) trees are highly synchronous at the individual tree level for AD 1900–2004 (Fig. 5.2a, 5.3a). The lack of climate-signal age-effects (Esper et al., 2008) as well as the robustness of the mean based on five trees are emphasized in the comparison with the isotope series obtained from four younger trees (Fig. 5.2d, 5.3d). The δ^{13}C and δ^{18}O series from both age classes yield a virtually identical mean. To assess this signal strength back in time, 51-year centered running RBAR and EPS (the latter is not shown in Fig. 5.2 and Fig. 5.3) statistics were computed based on the individual isotope values available every 10th year.

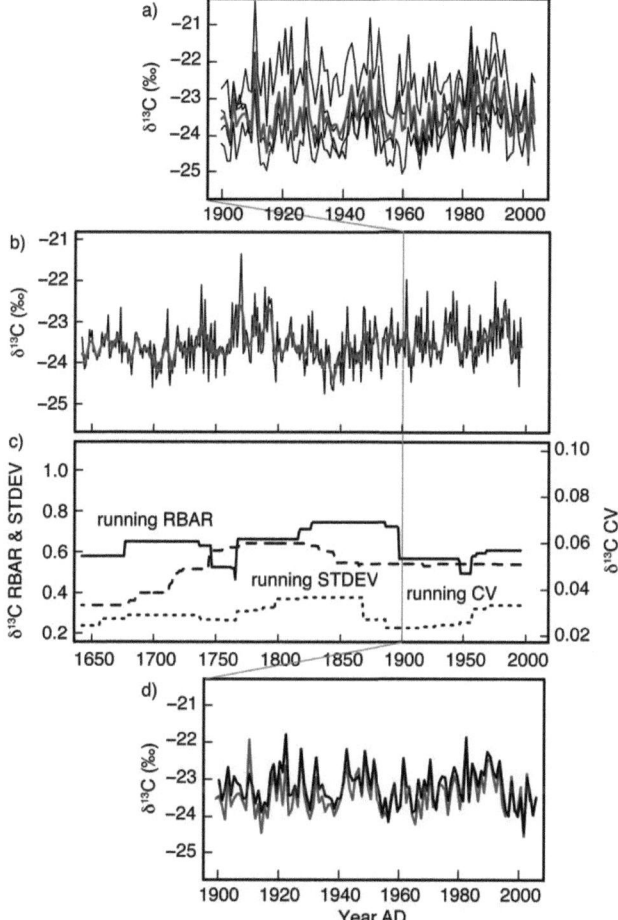

Figure 5.2: Carbon-chronology characteristics: (a) Single-tree series (black lines) and their average (red line) for AD 1900–2004; (b) annually resolved mean of five trees (black) and a 5-year centered running mean (red) for AD 1650–2004; c) 51-year moving windows of inter-series correlations (RBAR) and coefficient of variation (CV) of the split-pool values (every 10th year) as well as the standard deviation of the chronology mean (STDEV); d) comparison with younger larch trees (150–250 years; black line) for the 20th century.

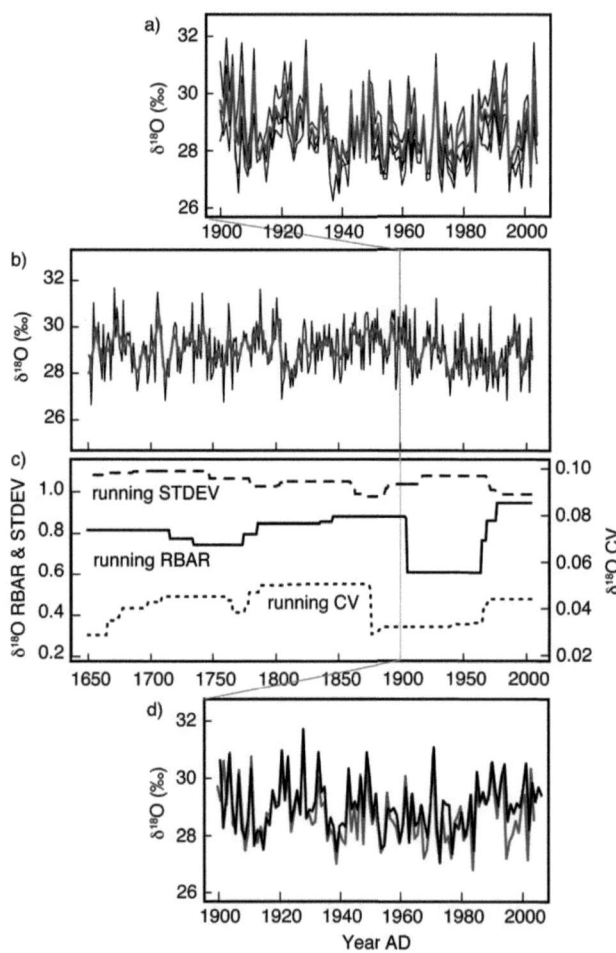

Figure 5.3: Oxygen-chronology characteristics (panels as in Fig. 5.2)

5.3 Results

Both isotope series showed a fairly high and stable RBAR (δ^{13}C: RBAR$_{mean}$ = 0.64; δ^{18}O: RBAR$_{mean}$ = 0.77) and exceed the EPS threshold of 0.85 throughout the entire period (δ^{13}C: EPS$_{mean}$ = 0.89; δ^{18}O: EPS$_{mean}$ = 0.94), pointing to a very strong common signal within the population and a robust mean chronology. For both isotope series the coefficient of variation (CV) showed a remarkably low dispersion between the single-tree isotope values, confirming not only a high synchronicity in the curve progression but also a narrow range of values in the isotope series of different trees.

In addition, the 51-year running standard deviation (STDEV) is illustrated in Figures 5.2 and 5.3 to investigate potential changes in variance in the chronologies (Frank et al., 2007b). While the STDEV of the oxygen series remained stable throughout the investigated period, the variance of the carbon series decreased slightly back in time. The consistently high RBAR, constant sample replication, and the small differences in the signal between the young and old-age trees (Fig. 5.2d, 5.3d) all minimize the likelihood for variance artifacts, but we are aware that changes in the variance of climatic time-series must be interpreted with caution(Frank et al., 2007b). The stability of the RBAR, EPS and CV statistics are particularly striking when considering that all calculations are limited to very few values (tree-wise analysis was only conducted every 10th year for AD 1650–1899 and simulated for AD 1900–2004). Overall, the signal strength assessment demonstrates a very high common signal, which is indicative of climatic forcing.

5.3.2 Climate signal(s)

For AD 1901–2004 bootstrapped correlation coefficients were calculated on a monthly basis for all δ^{13}C and δ^{18}O series and temperature, precipitation amount, sunshine duration, as well as the DRI (Fig. 5.4, 5.5). Significant correlations (p < 0.001), highlighted by colored bars, are restricted to a narrow window within the tree-line vegetation period, which is from mid May (stem growth starts mid June) to mid September (Moser et al., 2010), with highly significant positive correlations for temperature and sunshine duration and strong negative correlations with precipitation amount and DRI. All signs of the correlation coefficients agree with theoretical expectations based on isotope fractionation (McCarroll and Loader, 2004). No significant correlations were found with any month of the previous year (not shown).

Chapter 5 A 350-year drought reconstruction from Alpine tree-ring stable isotopes

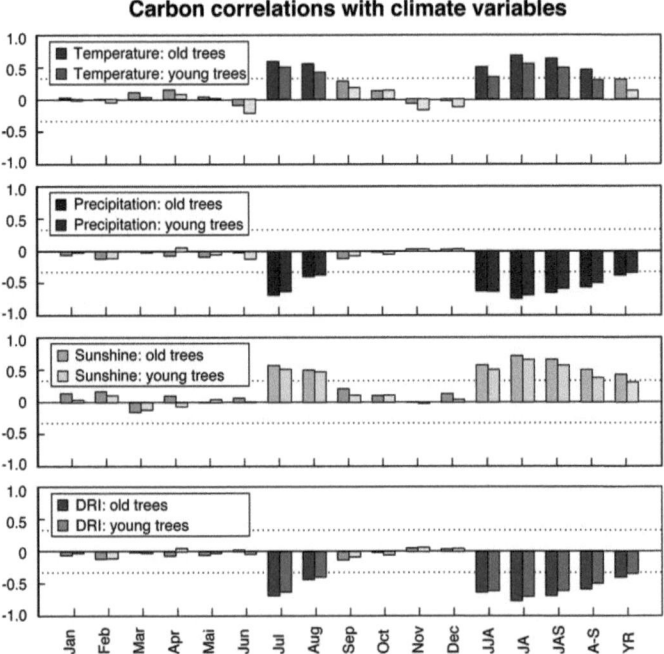

Figure 5.4: Colored bars illustrate significant bootstrapped correlation coefficients between δ^{13}C and temperature (red), precipitation amount (blue), sunshine duration (orange) and the calculated drought index (DRI, green) for AD 1901–2004. Dotted lines represent 99.9% confidence limits; grey bars represent insignificant correlations. All climate data were obtained from the Swiss part of the HISTALP-dataset, and all correlations were calculated for the old chronology trees (450–550 years; dark bars) and younger trees (150–250 years; light bars) from the same site.

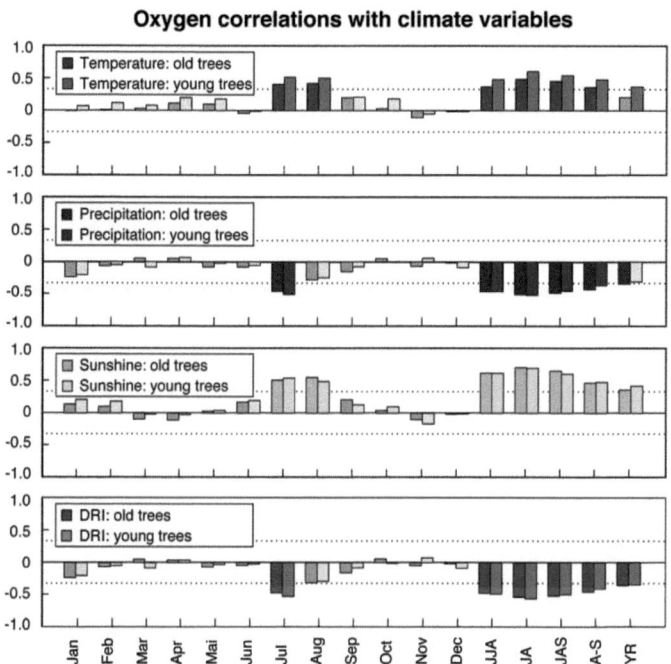

Figure 5.5: Correlation analysis of $\delta^{18}O$ and climate variables (panels as in Fig. 5.4)

Compared to climate response windows of tree-ring width and latewood density, this mid-late summer seasonal window appears particularly narrow and well defined. Highest correlation values were identified for both isotopes and all four climate variables when considering a July-August mean. Carbon isotopes correlate equally strong with all climate variables ($r \sim 0.7$). However, oxygen isotopes show a much stronger relationship with sunshine duration ($r \sim 0.7$) than with temperature, precipitation amount and DRI ($r \sim 0.5$). The strongest relationship was obtained between carbon and DRI ($r = -0.76$). As expected based upon the similarities in mean chronologies, all climate-isotope relationships are equally strong for the younger and older trees (Fig. 5.3). No distinct shifts in any of the climate-isotope relationships were found when the temporal stability of the climate isotope relationship was tested (data not shown and Kress et al., 2009a).

For central Europe strong spatial field correlations ($p < 0.001$) were found for both isotopes and both investigated climate variables (temperature and precipitation) with a particular emphasis on the Alpine arc (Fig. 5.6). While positive July-August temperature correlations (up to $r \sim 0.7$) emanated to southwestern Europe, the negative correlations with July-August precipitation amounts (up to $r \sim -0.7$) were restricted to central Europe. The climate-isotope relationships were somewhat more pronounced for carbon than for oxygen isotopes. Nevertheless, the spatial patterns were very similar for the two isotope series and the climate variables, suggesting a distinct regional climate signal in both isotope series.

5.3.3 Identification of the dominating climate signal in carbon isotopes

To assess the dominant climate signal in a long-term context, the carbon isotope series (AD 1650-2004) was compared to a July-August temperature and precipitation reconstruction from the corresponding grid cell of the Lötschental (AD 1659–2000, Casty et al., 2005) as well as to the July-August DRI calculated from these reconstructions (Fig. 5.7a). Over the 342-year period of overlap, the carbon isotope series showed high agreement with the temperature reconstruction and was anticorrelated with the precipitation record. Both relationships were stronger in the more recent period, where the reconstruction of Casty et al. (2005) consists of a dense network of high-quality

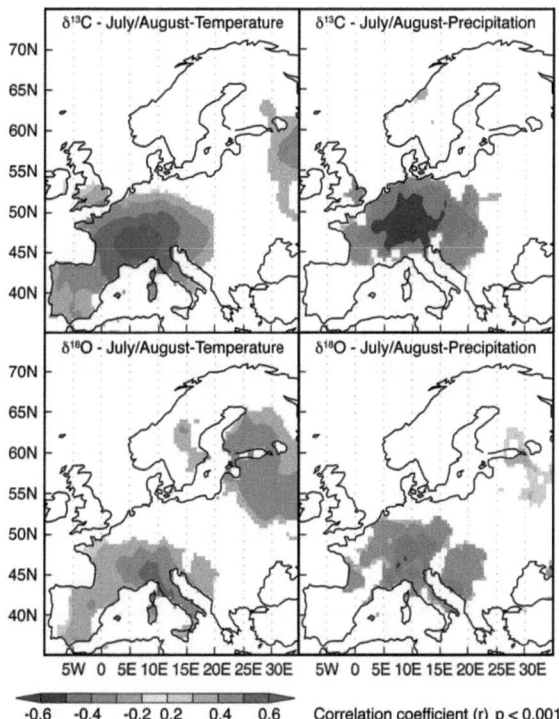

Figure 5.6: Significant spatial field correlations (p < 0.001) between δ^{13}C (top panels) and δ^{18}O (bottom panels) and July-August mean temperature and precipitation amount for AD 1901–2004. All climate data were obtained from a 0.5° x 0.5° monthly gridded meteorological dataset (CRU TS 3).

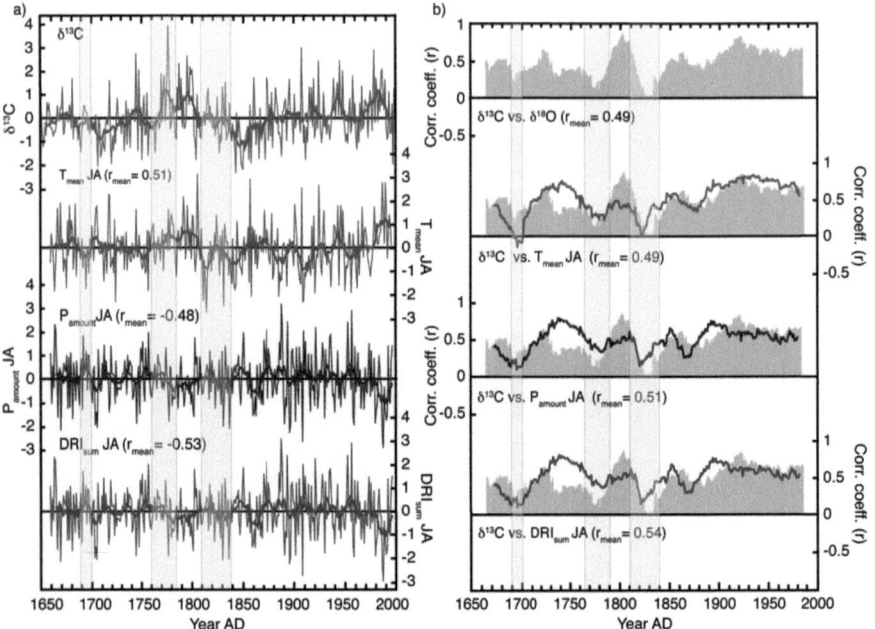

Figure 5.7: Comparison of $\delta^{13}C$ with 350-year climate reconstructions (Casty et al. 2005):

a) From top to bottom: $\delta^{13}C$ (blue), July-August temperature (T_{meanJA}; red), July-August precipitation ($P_{amountJA}$; dark blue) and July-August drought index (DRI_{sumJA}; green) calculated from T_{meanJA} and $P_{amountJA}$; bold lines represent 11-yr-centered running means, shaded areas demonstrate periods of weak correlations in b), and r-values (Pearson's r) indicate the mean correlations between annual $\delta^{13}C$-values and the corresponding climate variable. All values were standardized to the same mean (=0) and standard deviation (=1) (z-scores).

b) 31-yr-running correlations, centered, between $\delta^{13}C$ and $\delta^{18}O$ series (dark grey bars) compared to 31-yr-running correlations, centered, between $\delta^{13}C$ and T_{meanJA} (red), $P_{amountJA}$ (blue) and DRI_{sumJA} (green). The sign of correlation was inverted for $P_{amountJA}$ and DRI_{sumJA} data. A mean-between-series correlation is given by r. Shaded areas correspond to those in a) and indicate periods of weak correlations.

5.3 Results

instrumental data and over which the carbon isotope data were already screened.

The calculated DRI series is rather similar to the precipitation record but shows slightly stronger correlations with $\delta^{13}C$. To assess the temporal stability of the relationships, we compared 31-year running correlations of $\delta^{13}C$ with the three climate variables (Fig. 5.7b) as well as the relationship between $\delta^{13}C$ and $\delta^{18}O$. Most of the time a strong relationship between the two isotopes (r > 0.5) was evident. However, three periods with weaker correlations could be identified: 1690–1700, 1760–1785 and 1810–1840. Interestingly these periods of weak correlations between the two isotopes matched with weak correlations between carbon series and reconstructed temperature, precipitation, and DRI. In particular the running correlations between temperature and carbon showed a sharp decline within these periods, indicating that temperature alone may not be the dominating driving factor for the $\delta^{13}C$ signatures

To consider properties in the frequency domain, instrumental temperature, precipitation, calculated DRI and $\delta^{13}C$ series were analyzed using the multi-taper method (Mann and Lees, 1996) over the 1900–2004 period. While spectral power is similar in all variables at frequencies above ∼0.5 cycles/year, distinct differences become apparent in the lower frequency domain (Fig. 5.8), where the $\delta^{13}C$ spectrum meets neither the temperature nor the precipitation spectrum, but is rather located between the two. The DRI spectrum reveals a most similar balance of high to low frequency variance as the $\delta^{13}C$, and again is suggestive that the DRI, by combining temperature and precipitation effects, is the most suitable variable to explain $\delta^{13}C$ measurements. When the same analysis is conducted with the temperature and precipitation reconstructions by Casty et al. (2005) and compared to the spectrum of $\delta^{13}C$ for AD 1659–2000, the reconstructed temperatures have less low-frequency variability than the $\delta^{13}C$ (data not shown; see also paragraph above).

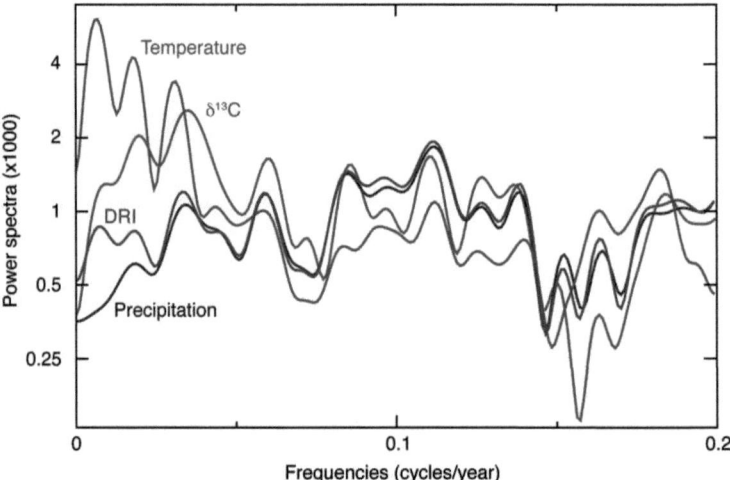

Figure 5.8: Multi-taper method (MTM) spectra using two-year resolution/ three tapers of the δ^{13}C (light blue), July-August temperature (red), July-August precipitation (dark blue) and the drought index (DRI, calculated from instrumental data, green) calculated over AD 1901–2004.

5.3.4 Drought reconstruction

Based on the time and frequency domain analyses, it became increasingly clear that the July-August DRI was likely the most appropriate reconstruction target. As a final test, relationships for the raw, high and low-passed δ^{13}C and DRI data (Fig. 5.9) over the 20th century yielded nearly identical correlations of -0.76, -0.75, and -0.81 for the raw, high and low-frequency domains. This suggests that the signal is appropriately balanced across all frequency domains and that δ^{13}C can likely be used to reconstruct inter-annual to at least multi-decadal signals with similar fidelity. Split-period verification of the calibrated relationships (Tab. 5.2) indicated a reliable model with high predictive skill by using ordinary least squares linear regression over the full 1900–2004 period.

Table 5.2: Calibration and verification statistics of δ^{13}C against July-August DRI using simple linear regression with R^2 (squared correlation coefficient), RMSE (root mean squared error), RE (reduction of error statistic), CE (coefficient of efficiency) and DW (Durbin-Watson statistic) for two independent periods.

	R^2	RSME	DW		R^2	RSME	RE	CE
1901–1952	0.62	0.65	2.55	1953–2004	0.54	0.66	0.60	0.60
1953–2004	0.54	0.63	2.30	1901–1952	0.62	0.65	0.52	0.52
Overall period (AD 1901-2004): $R^2 = 0.58$; DW = 2.49								

Chapter 5 A 350-year drought reconstruction from Alpine tree-ring stable isotopes

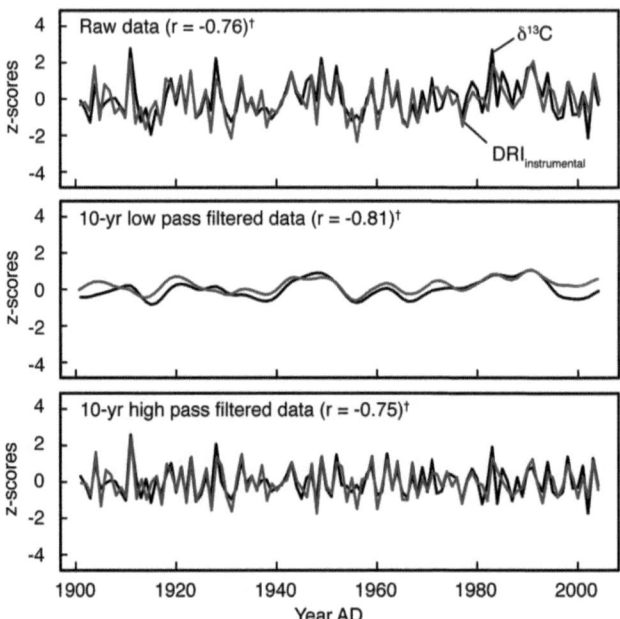

Figure 5.9: Carbon-isotope series (black) and July-August DRI calculated from instumental temperature an precipitation data (red) for AD 1901–2004 (top panel) and filtered (bottom panels) using cubic smoothing splines with 50% frequency-response cut-off at 10 years.
†The sign was inverted for all illustrated DRI-series while r (Pearson's r) is reporting the original corresponding relationship.

Our July-August DRI reconstruction (LOT$_{DRI}$; Fig. 5.10a) is presented with uncertainty estimates accounting for changes in the inter-series relationship (see Methods). The indicated extremes are the 20 driest and wettest summers throughout the reconstruction. The three driest summers were 1778, 1911 and 1983, the three wettest ones were 1845, 1852 and 1851. Considering the entire record, the late 17th and early 18th century were rather wet followed by a dry period in the later 18th century. The entire 19th century indicated wet conditions while, except for the very beginning, the

5.3 Results

20th century revealed rather dry conditions. On a decadal scale, four periods of severe droughts could be identified, centred on 1775, 1800, 1950 and 1990.

Table 5.3: Central European hydro-climatic series, which are compared to the new July-August drought reconstruction (LOT$_{DRI}$) in this study. The proportion of frequency variance was calculated for a 30-year high- (hp) and low-pass (lp) filtered series.

General information				Correlations with LOT$_{DRI}$ (r)		Proportion of frequency variance
Abbr.	Region	Signal/season	Source	raw	30yr lp	σ^2(30yr hp)/σ^2(30yr lp)
BUT	W-Carpathian	scPDSI/JJA	Büntgen et al. 2009	0.03	0.24	5.23
WMS	Vienna Basin	Precip./JJA	Wimmer (Büntgen et al. 2009)	0.20	0.43	6.75
WLR	Bavarian Forest	Precip/MA	Wilson et al. 2005	-0.04	-0.60	3.28
OBH	Austrian Alps	Precip/AMJ	Oberhuber and Kofler 2002	-0.03	-0.30	19.86

Chapter 5 A 350-year drought reconstruction from Alpine tree-ring stable isotopes

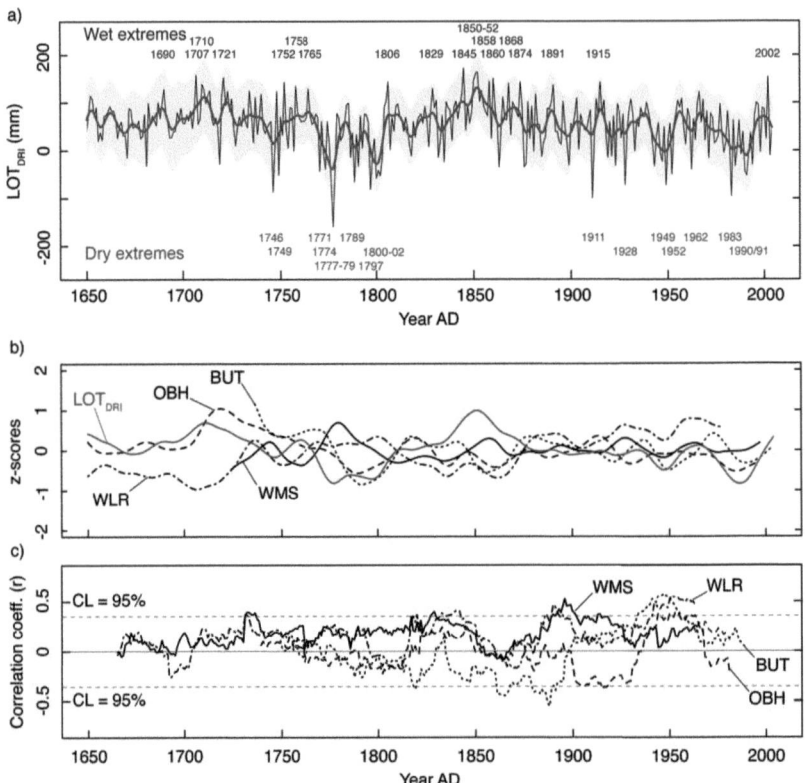

Figure 5.10: δ^{13}C drought reconstruction (LOT$_{DRI}$) and comparisons:

a) δ^{13}C drought reconstruction annually resolved and 10-year low pass filtered (bold line). Grey shadow indicates the ± RMSE/RBAR errormargin. The 20 wettest and driest years are indicated as wet/dry extreme events.

b) 30-year low pass filtered series of LOT$_{DRI}$ compared to one drought reconstruction (BUT) and three precipitation reconstructions (OBH, WMS, WLR) from different European regions. For details of these hydro-climatic reconstructions see Table 5.3.

c) 31-year centered running correlations between the annual LOT$_{DRI}$ and the other hydro-climatic records shown above (b) within a 95% confidence level (CL, bottom panel).

5.4 Discussion and Conclusions

Both the carbon and oxygen stable isotope series reveal extraordinarily strong and robust summer signals with all three investigated climatic variables (temperature, precipitation and sunshine duration). Key characteristics of the isotope-climate response include: i) highly significant correlations ($p < 0.001$) that were ii) stable through time, iii) regionally representative, and iv) evident in young and old trees. At the same time, our analyses failed to indicate a climatically dominating, and therefore primary controlling, factor for either of the isotopic variables. Including the results of the spectral analysis, it becomes apparent that $\delta^{13}C$ is best reflected by a combination of temperature and precipitation. Although $\delta^{18}O$ is likely also controlled by these two climate variables, it was not possible to increase the explained variance by a simple combination of them.

Although the climate-isotope relationships shown in Figure 5.4 and 5.5 are striking, they are just statistical relationships and may not be representative of the factors directly controlling isotope fractionation in plants. According to stable isotope theory, the main controls on isotope fractionation in trees are environmental factors such as stomatal conductance, assimilation rate, photon flux (for carbon), meteoric (source) water variability and leaf water enrichment (for oxygen) (McCarroll and Loader, 2004). These factors are externally influenced by insolation, leaf temperature, vapor pressure deficit and air mass properties, which, fortunately, are in turn closely correlated with meteorological variables such as air temperature, sunshine duration and precipitation amount (Loader et al., 2008). While at rather cool and moist high latitude sites, carbon isotopes in tree-rings tend to be dominated by variables that control the assimilation rate, mainly sunshine duration and air temperature (McCarroll and Pawellek, 2001; Gagen et al., 2007), at more xeric sites the climate signal in $\delta^{13}C$ signatures seems to be mainly related to stomatal conductance, which is controlled by air humidity and water availability and is therefore linked to antecedent precipitation (Leavitt and Long, 1988; Saurer et al., 1995). Our altitudinal treeline site in the Lötschental, although seemingly more similar to high latitude treeline conditions than to xeric sites, is situated in a transition zone between the oceanic moist regime of the outer parts of the Alps and the dry-subcontinental climate of the inner-alpine Rhône valley. It therefore is likely that in the Lötschental, $\delta^{13}C$ ratios are influenced by the interaction between assimilation rate and stomatal conductance. This is in agreement with Treydte et al.

Chapter 5 A 350-year drought reconstruction from Alpine tree-ring stable isotopes

(2001), who found equally strong climate correlations of temperature, precipitation and relative humidity with $\delta^{13}C$ from Lötschental spruce (*Picea abies*) and suggested a crucial influence of atmospheric humidity, and therefore stomatal conductance, on carbon signatures. Therefore we suggest that a simple drought index, as shown in this study, is more likely to capture the mixed climate signal in carbon isotopes than any individually measured climate variable(s).

Our study reveals higher correlations for temperature and precipitation with carbon than with oxygen values (Fig. 5.4, 5.5, 5.6), which is contrary to some recent studies showing equally strong climate correlations for both isotopes (Reynolds-Henne et al., 2007; Saurer et al., 2008) or stronger correlations for oxygen isotopes and summer temperature and precipitation (Raffalli-Delerce et al., 2004; Loader et al., 2008). While carbon isotopic ratios depend predominantly on leaf-internal processes, oxygen isotopic ratios in tree-ring cellulose reflect mostly a dampened leaf-water signal only (Yakir et al., 1990; Barbour et al., 2004), while they mainly record faithfully the isotopic composition of meteoric precipitation (Saurer et al., 1997b), which in turn depends strongly on the origin of the air masses (Dansgaard, 1964; Rozanski et al., 1993). Locations that receive air masses of the same origin during most of the time are thus likely to show a stronger climate-isotope relationship than locations that are characterized by a more complex synoptic situation. Northern European latitudes, which are strongly influenced by North Atlantic air masses, exhibit a particularly strong temperature sensitivity of $\delta^{18}O$ compared to other European sites (Treydte et al., 2007). The Alps, however, are characterized by a strong seasonal pattern: whilst influenced by Northern Atlantic Oscillation (NAO) in winter, the synoptic pattern in summer is much more complex with an increased influence of regionally to locally driven systems (Casty et al., 2005). This rather heterogeneous mixture of air mass sources during the vegetation period may influence the $\delta^{18}O$ signatures in the tree rings, resulting in weaker correlations with summer temperature and precipitation amount. Correlations with summer sunshine duration, however, remain fairly strong, as sunshine duration may reflect changes in the synoptic setting and atmospheric circulation that influence the $\delta^{18}O$ of precipitation. Longer sunshine hours will likely be accompanied by a decrease in precipitation and relative humidity and will therefore result in an increase of evaporative leaf water enrichment. The $\delta^{18}O$ signal should thus also contain water-balance information. In addition, sunshine duration is more closely related to the amount of shortwave radiation und thus leaf surface temperature, which also

5.4 Discussion and Conclusions

determines leaf $H_2^{18}O$ enrichment (Barbour et al., 2004).

Furthermore, the inability of carbon isotopes to capture the signal of one dominating climate variable seems to be linked to the relationships observed between carbon and oxygen isotopes (Fig. 5.7). While both isotopes are strongly correlated with each other most of the time, there are significant periods with almost no correlation, which coincide with periods of a weak summer temperature signal in carbon isotopes. In the lower frequency domain, DRI and $\delta^{13}C$ agreed well during the more recent period. Before AD 1850 in contrast (Fig. 5.7), the calculated DRI showed a reduction in low-frequency variance not seen in the $\delta^{13}C$. It should be considered that the DRI derived from the Casty et al. (2005) reconstruction becomes more uncertain back in time, changing from instrumental records in the modern period to fewer records overall and an increasing proportion of documentary and other, quantitatively less reliable sources. Also, methodological details may result in variance artifacts and losses (Frank et al., 2007b; Lee et al., 2008). The majority of instrumental records drops out around AD 1850, which coincides with the first and strongest period of weak correlations. In addition, the necessary homogenization procedures applied to instrumental station data are prone to error or uncertainty (Frank et al., 2007a). Nevertheless, DRI accounts for more variance in the $\delta^{13}C$ series than the temperature or precipitation reconstructions alone, and it also tends to have a more constant relationship over time. The fact that the strength of the $\delta^{18}O$–$\delta^{13}C$ and $\delta^{13}C$–DRI relationships generally rise and fall in parallel suggests linkages either to changes in the regional radiation budget (e.g., cloud cover) or to changes in the relationship between the climate variables. External factors, in particular moisture conditions, can influence both isotopic ratios (Saurer et al., 1997a). On the leaf level, for example, stomatal conductance has a major influence on both isotopes: carbon isotopic ratios are directly linked to stomatal conductance (fractionation due to diffusion), whereas oxygen isotopic ratios are influenced by transpiration (Yakir et al., 1990; Barbour et al., 2004), which is in turn coupled to stomatal conductance and source water. Drier conditions will therefore increase the values of both isotopes. This relationship between the two isotopes is expressed as positive correlations as long as other factors, which can alternate the isotopic composition, do not vary in time (Saurer et al., 1997a). The Alps, however, are situated within a varying influence of NAO (Casty et al., 2005). Air masses can thus originate from different sources resulting in varying $\delta^{18}O$ values in precipitation, but they may also affect the temperature-precipitation relationship and therefore $\delta^{13}C$ (Reynolds-Henne et al., 2007).

The DRI explains more variance in $\delta^{13}C$ than temperature or precipitation alone, which is expressed in manifold ways: (i) carbon isotopes correlate more strongly with DRI than any other climate variable (Fig. 5.4, 5.5, 5.6), (ii) this response is consistent for the high-pass, low-pass, and unfiltered data (Fig. 5.9), (iii) the spectral power is similar across all frequency domains (Fig. 5.8), and (iv) calibration and verification statistics are strong (Tab. 5.2). These relationships and properties result in a reliable and robust DRI reconstruction (LOT_{DRI}) with error estimates that account for changes in the inter-series correlation of the individual chronology trees and are therefore of varying bandwidth. Due to the preservation of the low frequency signal, the record not only reveals information about the occurrence of extremes, but is also able to capture decadal-scale variability. Interestingly, extremely dry summers occur only during two periods, namely during the second half of the 18th century and during the 20th century.

The LOT_{DRI} reconstructed drought history was compared with other tree-ring based reconstructions available for central Europe, by considering relationships of the raw data, after low-pass filtering, and through time using running correlations (Fig. 5.10b,c and Tab. 5.3). The only record showing a significant positive correlation with LOT_{DRI} at all investigated frequencies is the precipitation reconstruction by Wimmer et al. (WMS, Büntgen et al., 2009) for the Vienna basin. If WMS and LOT_{DRI} are compared in more detail, some similarities can be found: a rather wet late 17th and early 18th century, and a dry 20th century, while the other records do not show consistent correlations in the raw data. If 30-year low-pass filtered series are compared, the precipitation record of the Bavarian forest (WLR, Wilson et al., 2005) seem to be rather anticorrelated to the LOT_{DRI} record (r = -0.60), while the precipitation record of the Austrian Alps (OBH, Oberhuber and Kofler, 2002) and the Slovakian drought record (BUT, Büntgen et al., 2009) express rather weak relationships to our LOT_{DRI} reconstruction in the lower-frequency domain. This may also indicate different preservation of low frequency in the records. While the proportion of 30-year high to low frequency variance for LOT_{DRI} (4.41) is in the range of BUT, WLR and WMS, the OBH (19.86; compare Tab. 5.3) seems to contain hardly any lower-frequency information. Nevertheless, regarding the spatial field correlations of $\delta^{13}C$ and precipitation (Fig. 5.6), which diminish towards the North and East, the regional nature of precipitation variability and the rather large seasonal windows (in particular for OBH and WLR) compared to the very narrow and distinct response window of the LOT_{DRI}

5.4 Discussion and Conclusions

(Fig. 5.4, 5.5), it is rather remarkable that some coherence can be found between these hydroclimatic records and the LOT$_{DRI}$.

We chose to reconstruct DRI; however, other drought metrics such as the self-calibrating Palmer Drought Severity Index (scPDSI) are also widely used (Dai et al., 2004; van der Schrier et al., 2006, 2007). A comparison of all scPDSI records during the 20th century for the grid-cell closest to the Lötschental (Tab. 5.4) shows weak and insignificant correlations between the record of the Alps (van der Schrier et al., 2007) and the corresponding gridpoint from the European (van der Schrier et al., 2006), or global (Dai et al., 2004) datasets. However, the latter two datasets agree reasonably well at this location. Despite this inhomogeneous pattern of scPDSI series, the DRI shows significant, although rather weak correlations with all three of them (Tab. 5.4).

Table 5.4: AD1901–2004 correlation matrix between DRI (calculated) and three datasets of scPDSI from the grid cell closest to the Lötschental (***p < 0.001).

	DRI	scPDSI Europe	scPDSI Alps
scPDSI Europe (van der Schrier et al., 2006)	0.45***		
scPDSI Alps (van der Schrier et al., 2007)	0.49***	0.10	
scPDSI Globe (Dai et al., 2004)	0.38***	0.61***	0.05

This may be caused by the different approaches in calculating the scPDSI and the DRI. Although the scPDSI takes snowmelt into account, its calculation involves a rather complex water budget system, which contains empirically weighted terms of previous months' conditions and accounts therefore for a memory effect of the preceding months (van der Schrier et al., 2006). In contrast, DRI was calculated only based upon the current month's temperature and precipitation data and thus does not account for any storage effects of the previous months. Despite these differences in concept and practice, some commonalities are still observed even at the continental-scale. For example, considering a mean of scPDSI observations for the European continent for AD 1900–2004 (van der Schrier et al., 2006), the wettest summer, AD 1915, is well replicated in the LOT$_{DRI}$ reconstruction representing the 2nd wettest summer in the 20th century. Similarly, the driest summer of the scPDSI series, AD 1947, is recognized as drier than average in the LOT$_{DRI}$. When this mean European summer scPDSI is

Chapter 5 A 350-year drought reconstruction from Alpine tree-ring stable isotopes

compared to the LOT_{DRI} on a decadal scale, all wet periods and all but one dry periods are common to both records. This coherence between a European mean of scPDSI data and the LOT_{DRI} again emphasizes the regionally extensive signal captured by the isotopic data from the Lötschental (Fig. 5.6).

Overall, our study shows that climate reconstructions based on isotopic data using simple regression models work reasonably well as long as the relationships between meteorological variables and environmental factors controlling the isotope fractionation remain stable in time. As the Alps are situated within a varying influence of the NAO (Casty et al., 2005), we cannot expect a stable temperature-precipitation relationship back in time. Oxygen isotopes may most strongly reflect sunshine duration as meteoric variable because this is an indirect measure of evaporative leaf water enrichment. Furthermore, the analysis of the relationship between carbon and oxygen isotopes may contribute to identify instabilities between meteorological variables for periods without instrumental weather data, as the different fractionation processes are not automatically linked to the same climate variables. Indices that combine temperature and moisture influences will account for instabilities between these meteoric variables, and by accounting for the different mechanisms controlling isotopic fractionation, will yield more reliable climate reconstructions. Accordingly, we presented a regional summer drought reconstruction that provides new evidence for summer temperature and water availability back to AD 1650 in the Alps, and a small, yet crucial step towards a comprehensive understanding of long-term changes in Europe's hydroclimate.

Acknowledgments

This work was funded by the EU project FP6-2004-GLOBAL-017008-2 (MILLENNIUM). JE and DCF acknowledge support from the Swiss National Science Foundation (NCCR-Climate). We thank W. Oberhuber, R. Wimmer and R. Wilson for making their reconstructions available and in particular to U. Büntgen, who provided his very recent Slovakian drought reconstruction. Many thanks to K. Treydte, A. Verstege, D. Nievergelt, M. Tröndle and L. Läubli for helpful discussions and technical support and to C. Bigler and P. Weibel for sharing their rich experience in drought indices.

References

Auer, I., Böhm, R., Jurkovic, A., Lipa, W., Orlik, A., Potzmann, R., Schoner, W., Ungersbock, M., Matulla, C., Briffa, K., Jones, P., Efthymiadis, D., Brunetti, M., Nanni, T., Maugeri, M., Mercalli, L., Mestre, O., Moisselin, J. M., Begert, M., Muller-Westermeier, G., Kveton, V., Bochnicek, O., Stastny, P., Lapin, M., Szalai, S., Szentimrey, T., Cegnar, T., Dolinar, M., Gajic-Capka, M., Zaninovic, K., Majstorovic, Z., and Nieplova, E. (2007). HISTALP - historical instrumental climatological surface time series of the Greater Alpine Region. *International Journal of Climatology*, 27(1):17–46; doi:10.1002/joe.1377.

Barbour, M. M., Roden, J. S., Farquhar, G. D., and Ehleringer, J. R. (2004). Expressing leaf water and cellulose oxygen isotope ratios as enrichment above source water reveals evidence of a Péclet effect. *Oecologia*, 138(3):426–435; doi:10.1007/s00442-003-1449-3.

Bigler, C., Braker, O. U., Bugmann, H., Dobbertin, M., and Rigling, A. (2006). Drought as an inciting mortality factor in Scots pine stands of the Valais, Switzerland. *Ecosystems*, 9(3):330–343; doi:10.1007/s10021-005-0126-2.

Blasing, T., Duvick, D., and West, D. (1981). Dendroclimatic calibration and verification using regionally averaged and single station precipitation data. *Tree-Ring Bulletin*, 41:37–43.

Boda, S. Y., Treydte, K. S., Fonti, P., Gessler, A., Graf-Pannatier, E., Saurer, M., Siegwolf, R. T. W., and Werner, W. (subm.). Intra-seasonal pathway of oxygen isotopes from soil to wood in the Loetschental (Swiss Alps). *Plant, Cell and Environment*.

Boettger, T., Haupt, M., Knoller, K., Weise, S. M., Waterhouse, J. S., Rinne, K. T., Loader, N. J., Sonninen, E., Jungner, H., Masson-Delmotte, V., Stievenard, M., Guillemin, M. T., Pierre, M., Pazdur, A., Leuenberger, M., Filot, M., Saurer, M., Reynolds, C. E., Helle, G., and Schleser, G. H. (2007). Wood cellulose preparation methods and mass spectrometric analyses of $\delta^{13}C$, $\delta^{18}O$ and nonexchangeable $\delta^{2}H$ values in cellulose, sugar, and starch: an interlaboratory comparison. *Analytical Chemistry*, 79(12):4603–4612; doi:10.1021/ac0700023.

Brazdil, R., Pfister, C., Wanner, H., Von Storch, H., and Luterbacher, J. (2005).

Historical climatology in Europe - the state of the art. *Climatic Change*, 70(3):363–430; doi:10.1007/s10584-005-5924-1.

Bugmann, H. and Cramer, W. (1998). Improving the behaviour of forest gap models along drought gradients. *Forest Ecology and Management*, 103(2-3):247–263; doi:10.1016/S0378-1127(97)00217-X.

Büntgen, U., Brazdil, R., Frank, D., and Esper, J. (2009). Three centuries of Slovakian drought dynamics. *Climate Dynamics*, pages in press; doi:10.1007/s00382-009-0563-2.

Büntgen, U., Esper, J., Frank, D. C., Nicolussi, K., and Schmidhalter, M. (2005). A 1052-year tree-ring proxy for Alpine summer temperatures. *Climate Dynamics*, 25(2-3):141–153; doi:10.1007/s00382-005-0028-1.

Büntgen, U., Frank, D. C., Niervergelt, D., and Esper, J. (2006). Summer temperature variations in the European Alps, AD 755-2004. *Journal of Climate*, 19(21):5606–5623; doi:10.1175/JCLI3917.1.

Casty, C., Wanner, H., Luterbacher, J., Esper, J., and Böhm, R. (2005). Temperature and precipitation variability in the European Alps since 1500. *International Journal of Climatology*, 25:1855–1880; doi:10.1002/joc.1216.

Cook, E. R., Briffa, K. R., and Jones, P. D. (1994). Spatial regression methods in dendroclimatology: a review and comparison of two techniques. *International Journal of Climatology*, 14(4):379–402; doi:10.1002/joc.3370140404.

Cook, E. R., Briffa, K. R., Meko, D. M., Graybill, D. A., and Funkhouser, G. (1995). The segment length curse in long tree-ring chronology development for paleoclimatic studies. *The Holocene*, 5(2):229–237; doi:10.1177/095968369500500211.

Cook, E. R. and Peters, K. (1981). The smoothing spline: a new approach to standardizing forest interior tree-ring width series for dendroclimatic studies. *Tree Ring Bulletin*, 41:45–53.

Cook, E. R., Woodhouse, C. A., Eakin, C. M., Meko, D. M., and Stahle, D. W. (2004). Long-term aridity changes in the western United States. *Science*, 306(5698):1015–1018; doi:10.1126/science.1102586.

Dai, A., Trenberth, K., and Qian, T. (2004). A global dataset of Palmer Drought

5.4 Discussion and Conclusions

Severity Index for 1870–2002: relationship with soil moisture and effects of surface warming. *Journal of Hydrometeorology*, 5:1117–1130; doi:10.1175/JHM-386.1.

Dansgaard, W. (1964). Stable isoptopes in precipitation. *Tellus*, 16:436–468.

Durbin, J. amd Watson, G. (1951). Testing for serial correlation in least squares regression. *Biometrika*, 38:159–178.

Esper, J., Büntgen, U., Frank, D. C., Niervergelt, D., and Liebhold, A. (2007a). 1200 years of regular outbreaks in alpine insects. *Proceedings of the Royal Society B*, 274:671–679; doi:10.1098/rspb.2006.0191.

Esper, J., Frank, D., Buntgen, U., Verstege, A., and Luterbacher, J. (2007b). Long-term drought severity variations in Morocco. *Geophysical Research Letters*, 34(17):L17702; doi:10.1029/2007GL030844.

Esper, J., Niederer, R., Bebi, P., and Frank, D. (2008). Climate signal age effects-evidence from young and old trees in the Swiss Engadin. *Forest Ecology and Management*, 255(11):3783–3789; doi:10.1016/j.foreco.2008.03.015.

Farquhar, G. D., Ehleringer, J. R., and Hubick, K. T. (1989). Carbon isotope discrimination and photosynthesis. *Annual Review of Plant Physiology and Plant Molecular Biology*, 40:503–537; doi:10.1146/annurev.pp.40.060189.002443.

Frank, D., Buntgen, U., Bohm, R., Maugeri, M., and Esper, J. (2007a). Warmer early instrumental measurements versus colder reconstructed temperatures: shooting at a moving target. *Quaternary Science Reviews*, 26(25-28):3298–3310; doi:10.1016/j.quascirev.2007.08.002.

Frank, D. and Esper, J. (2005). Characterisazion and climate response patterns of a high-elevation multi-species tree-ring network in the European Alps. *Dendrochronologia*, 22:107–121; doi:10.1016/j.dendro.2005.02.004.

Frank, D., Esper, J., and Cook, E. R. (2007b). Adjustment for proxy number and coherence in a large-scale temperature reconstruction. *Geophysical Research Letters*, 34(16):L16709; doi:10.1029/2007GL030571.

Friedrichs, D. A., Buntgen, U., Frank, D. C., Esper, J., Neuwirth, B., and Loffler, J. (2009). Complex climate controls on 20th century oak growth in central-west Germany. *Tree Physiology*, 29(1):39–51; doi:10.1093/treephys/tpn003.

Fritts, H. C. (1976). *Tree Rings and Climate*. Academic Press, London, England.

Gagen, M., McCarroll, D., Loader, N. J., Robertson, L., Jalkanen, R., and Anchukaitis, K. J. (2007). Exorcising the 'segment length curse': summer temperature reconstruction since AD 1640 using non-detrended stable carbon isotope ratios from pine trees in northern Finland. *The Holocene*, 17(4):435–446; doi:10.1177/0959683607077012.

Ghil, M., Allen, M. R., Dettinger, M. D., Ide, K., Kondrashov, D., Mann, M. E., Robertson, A. W., Saunders, A., Tian, Y., Varadi, F., and Yiou, P. (2002). Advanced spectral methods for climatic time series. *Reviews of Geophysics*, 40(1):doi:10.1029/2000RG000092.

Guiot, J. (1991). The bootstrapped response function. *Tree-Ring Bulletin*, 51:39–41.

Hilasvuori, E., Berninger, F., Sonninen, E., Tuomenvirta, H., and Jungner, H. (2009). Stability of climate signal in carbon and oxygen isotope records and ring width from Scots pine (*Pinus sylvestris* L.) in Finland. *Journal of Quaternary Science*, 24(5):469–480; doi:10.1002/jqs.1260.

Holmes, R. L. (1983). Computer-assisted quality control in tree-ring dating and measurements. *Tree-Ring Bulletin*, 43:69–78.

Kress, A., Saurer, M., Büntgen, U., Treydte, K., Bugmann, H., and Siegwolf, R. T. W. (2009a). Summer temperature dependency of larch budmoth outbreaks revealed by Alpine tree-ring isotope chronologies. *Oecologia*, 160(2):353–365; doi:10.1007/s00442-009-1290-4.

Kress, A., Young, G. H. F., Saurer, M., Loader, N. J., Siegwolf, R. T. W., and McCarroll, D. (2009b). Stable isotope coherence in the earlywood and latewood of tree-line conifers. *Chemical Geology*, 268(1-2):52–57; doi:10.1016/j.chemgeo.2009.07.008.

Leavitt, S. W. and Long, A. (1988). Stable carbon isotope chronologies from trees in the southwestern United States. *Global Biogeochemical Cycles*, 2(3):189–198; doi:10.1029/GB002i003p00189.

Lee, T., Zwiers, F., and Tsao, M. (2008). Evaluation of proxy-based millennial reconstruction methods. *Climate Dynamics*, 31(2):263–281; doi:10.1007/s00382-007-0351-9.

Leuenberger, M. (2007). To what extent can ice core data contribute to the understanding of plant ecological developments of the past? In Dawson, T. and Siegwolf, R., editors, *Stable Isotopes as Indicators of Ecological Change*, pages 211–233. Elsevier Academic Press, London.

Loader, N. J., Santillo, P. M., Woodman-Ralph, J. P., Rolfe, J. E., Hall, M. A., Gagen, M., Robertson, I., Wilson, R., Froyd, C. A., and McCarroll, D. (2008). Multiple stable isotopes from oak trees in southwestern Scotland and the potential for stable isotope dendroclimatology in the maritime climatic regions. *Chemical Geology*, 252:62–71; doi:10.1016/j.chemgeo.2008.01.006.

Mann, M. E. and Lees, J. (1996). Robust estimation of background noise and signal detection in climatic time series. *Climatic Change*, 33:409–445; doi:10.1007/BF00142586.

McCarroll, D. and Loader, N. J. (2004). Stable isotopes in tree rings. *Quaternary Science Reviews*, 23(7-8):771–801; doi:10.1016/j.quascirev.2003.06.017.

McCarroll, D. and Pawellek, F. (2001). Stable carbon isotope ratios of *Pinus sylvestris* form northern Finland and the potential for extracting a climate signal from long Fennoscandian chronologies. *The Holocene*, 11(5):517–526; doi:10.1191/095968301680223477.

Mitchell, T. D. and Jones, P. D. (2005). An improved method of constructing a database of monthly climate observations and associated high-resolution grids. *International Journal of Climatology*, 25(6):693–712; doi:10.1002/joc.1181.

Moser, L., Fonti, P., Büntgen, U., Esper, J., Luterbacher, J., Franzen, J., and Frank, D. (2010). Timing and duration of European larch growing season along an altitudinal gradient in the Swiss Alps. *Tree Physiology*, 30(2):225–233; doi:10.1093/treephys/tpp108.

Oberhuber, W. and Kofler, W. (2002). Dendroclimatological spring rainfall reconstruction for an inner Alpine dry valley. *Theoretical and Applied Climatology*, 71:97–106; doi:10.1007/s704-002-8210-8.

Pauling, A., Luterbacher, J., Casty, C., and Wanner, H. (2006). Five hundred years of gridded high-resolution precipitation reconstructions over Europe and the connection to large-scale circulation. *Climate Dynamics*, 26(4):387–405; doi:10.1007/s00382-005-0090-8.

Paulsen, J. and Korner, C. (2001). GIS-analysis of tree-line elevation in the Swiss Alps suggests no exposure effect. *Journal of Vegetation Science*, 12(6):817–824; doi:10.2307/3236869.

Raffalli-Delerce, G., Masson-Delmotte, V., Dupouey, J. L., Stievenard, M., Breda, N., and Moisselin, J. M. (2004). Reconstruction of summer droughts using tree-ring cellulose isotopes: a calibration study with living oaks from Brittany (western France). *Tellus B*, 56(2):160–174; doi:10.1111/j.1600–0889.2004.00086.x.

Raible, C. C., Casty, C., Luterbacher, J., Pauling, A., Esper, J., Frank, D. C., Buntgen, U., Roesch, A. C., Tschuck, P., Wild, M., Vidale, P. L., Schar, C., and Wanner, H. (2006). Climate variability-observations, reconstructions, and model simulations for the Atlantic-European and Alpine region from 1500-2100 AD. *Climatic Change*, 79(1-2):9–29; doi:10.1007/s10584–006–9061–2.

Reynolds-Henne, C. E., Siegwolf, R. T. W., Treydte, K. S., Esper, J., Henne, S., and Saurer, M. (2007). Temporal stability of climate-isotope relationships in tree rings of oak and pine (Ticino, Switzerland). *Global Biogeochemical Cycles*, 21(4):GB4009; doi:10.1029/2007GB002945.

Roden, J. S., Lin, G., and Ehleringer, J. R. (2000). A mechanistic model for interpretation of hydrogen and oxygen isotope ratios in tree-ring cellulose. *Geochimica et Cosmochimica Acta*, 64(1):21–35; doi:10.1016/S0016–7037(99)00195–7.

Rozanski, K., Arguas-Arguas, L., and Gonfiantini, R. (1993). Isotopic patterns in modern global precipitation. In Swart, P., editor, *Climate Change in Continental Isotopic Records*, volume 78 of *Geophysical Monograph*, pages 1–36. American Geophysical Union, Washington, DC.

Saurer, M., Aellen, K., and Siegwolf, R. (1997a). Correlating $\delta^{13}C$ and $\delta^{18}O$ in cellulose of trees. *Plant, Cell and Environment*, 20(12):1543–1550; doi:10.1111/j.1365–3040.1997.tb00733.x.

Saurer, M., Borella, S., and Leuenberger, M. (1997b). $\delta^{18}O$ of tree rings of beech (*Fagus sylvatica*) as a record of $\delta^{18}O$ of the growing season precipitation. *Tellus B*, 49B:80–92; doi:10.1034/j.1600–0889.49.issue1.6.x.

Saurer, M., Cherubini, P., Reynolds-Henne, C. E., Treydte, K. S., Anderson, W. T., and Siegwolf, R. T. W. (2008). An investigation of the common signal in tree ring stable isotope chronologies at temperate sites. *Journal of Geophysical Research*, 113:G04035; doi:10.1029/2008JG000689.

Saurer, M., Siegenthaler, U., and Schweingruber, F. H. (1995). The climate-carbon isotope relationship in tree-rings and the significance of site conditions. *Tellus B*, 47(3):320–330; doi:10.1034/j.1600–0889.47.issue3.4.x.

Saurer, M. and Siegwolf, R. (2004). Pyrolysis techniques for oxygen isotope analysis of cellulose. In *Handbook of Stable Isotope Analytical Techniques*, volume 1, pages 497–508; doi:10.1016/B978–044451114–0/50025–9. Elsevier, New York.

Seager, R., Graham, N., Herweijer, C., Gordon, A. L., Kushnir, Y., and Cook, E. (2007). Blueprints for medieval hydroclimate. *Quaternary Science Reviews*, 26(19-21):2322–2336; doi:10.1016/j.quascirev.2007.04.020.

Stokes, M. A. and Smiley, T. L. (1968). *An introduction to tree-ring dating*. (reprinted 1996). University of Arizona Press, Tucson, US, Chicago.

Thornthwaite, C. (1948). An approach toward a rational classification of climate. *Geographical Review*, 38:55–94; doi:10.2307/210739.

Touchan, R., Anchukaitis, K. J., Meko, D. M., Attalah, S., Baisan, C., and Aloui, A. (2008). Long term context for recent drought in northwestern Africa. *Geophysical Research Letters*, 35(13):L13705; doi:10.1029/2008GL034264.

Treydte, K., Frank, D., Esper, J., Andreu, L., Bednarz, Z., Berninger, F., Boettger, T., D'Alessandro, C. M., Etien, N., Filot, M., Grabner, M., Guillemin, M. T., Gutierrez, E., Haupt, M., Helle, G., Hilasvuori, E., Jungner, H., Kalela-Brundin, M., Krapiec, M., Leuenberger, M., Loader, N. J., Masson-Delmotte, V., Pazdur, A., Pawelczyk, S., Pierre, M., Planells, O., Pukiene, R., Reynolds-Henne, C. E., Rinne, K. T., Saracino, A., Saurer, M., Sonninen, E., Stievenard, M., Switsur, V. R., Szczepanek, M., Szychowska-Krapiec, E., Todaro, L., Waterhouse, J. S., Weigl, M., and Schleser, G. H. (2007). Signal strength and climate calibration of a European tree-ring isotope network. *Geophysical Research Letters*, 34(24):L24302; doi:10.1029/2007GL031106.

Treydte, K. S., Schleser, G. H., Helle, G., Frank, D. C., Winiger, M., Haug, G. H., and Esper, J. (2006). The twentieth century was the wettest period in northern Pakistan over the past millennium. *Nature*, 440(7088):1179–1182; doi:10.1038/nature04743.

Treydte, K. S., Schleser, G. H., Schweingruber, F. H., and Winiger, M. (2001). The climatic significance of $\delta^{13}C$ in subalpine spruces (Lötschental, Swiss Alps). *Tellus B*, 53(5):593–611; doi:10.1034/j.1600-0889.2001.530505.x.

van der Schrier, G., Briffa, K. R., Jones, P. D., and Osborn, T. J. (2006). Summer moisture variability across Europe. *Journal of Climate*, 19(12):2818–2834; doi:10.1175/JCLI3734.1.

van der Schrier, G., Efthymiadis, D., Briffa, K. R., and Jones, P. D. (2007). European Alpine moisture variability for 1800-2003. *International Journal of Climatology*, 27(4):415–427; doi:10.1002/joc.1411.

Wigley, T. M. L., Briffa, K. R., and Jones, P. D. (1984). On the average value of correlated time series, with applications in dendroclimatology and hydrometeorology. *Journal of Applied Meteorology*, 23(2):201–213; doi:10.1175/1520-0450(1984)023<0201:OTAVOC>2.0.CO;2.

Wilks, D. (2006). *Statistical methods in the atmospheric sciences*. Academic Press, 2nd edition.

Wilson, R. J. S., Luckmann, B. H., and Esper, J. (2005). A 500 year dendroclimatic reconstruction of spring-summer precipitation from the lower Bavarian Forest region, Germany. *International Journal of Climatology*, 25(5):611–630; doi:10.1002/joc.1150.

Yakir, D., DeNiro, M. J., and Ephrath, J. E. (1990). Effects of water-stress on oxygen, hydrogen and carbon isotope ratios in two species of cotton plants. *Plant, Cell and Environment*, 13(9):949–955; doi:10.1111/j.1365-3040.1990.tb01985.x.

Young, G. H. F., McCarroll, D., Loader, N. J., and Kirchhefer, A. J. (2010). 500-years of summer near-ground solar radiation from stable carbon isotopes in Norwegian tree-rings. *The Holocene*, 20(3):315–324; doi:10.1177/0959683609351902.

6

1200 years of climate history from stable carbon and oxygen isotopes

Anne Kress[1], Matthias Saurer[1], Sarah Hangartner[2], Markus Leuenberger[2], Ulf Büntgen[3], David C. Frank[3], Rolf T.W. Siegwolf[1], Urs Baltensperger[1], and Harald Bugmann[4*]

[1] *Laboratory of Atmospheric Chemistry, Paul Scherrer Institut, CH-5232 Villigen PSI, Switzerland*
[2] *Climate and Environmental Physics, University of Bern, CH-3012 Bern, Switzerland*
[3] *Swiss Federal Research Institute WSL, CH-8903 Birmensdorf, Switzerland*
[4] *Forest Ecology, Department of Environmental Sciences, Swiss Federal Institute of Technology Zürich, CH-8092 Zürich, Switzerland*

Manuscript in preparation

―――――――――――

*Preliminary author list

Chapter 6 1200 years of climate history from stable carbon and oxygen isotopes

6.1 Introduction

Twentieth-century warming is accompanied by an altering temporal and spatial distribution of precipitation (Jansen et al., 2007). Such hydrologic changes can affect human well-being and ecosystem dynamics more strongly than the rising temperature itself (Kundzewicz et al., 2007). However, while the past, present and projected future rates of regional to global temperature have been extensively investigated, only little is known about past changes in precipitation variability and the hydrological cycle (Huntington, 2006; Seager et al., 2007). Climate reconstructions addressing past moisture conditions are vastly underrepresented especially in central Europe (Büntgen et al., 2009; Kress et al., 2010).

This lack of moisture-sensitive records is accompanied by persisting controversy surrounding the magnitude, extent and nature of major climate episodes, such as the Medieval Warm Period (MWP; e.g., Hughes and Diaz, 1994) and the Little Ice Age (LIA; e.g., Overpeck et al., 1997) on a global scale. In particular, the onset and magnitude of MWP is an object of ongoing discussion since its first description more than 40 years ago (Lamb, 1965). While Hughes and Diaz (1994) concluded that the available evidence does not support a global MWP, Esper et al. (2005) found that comparisons of large scale temperature reconstructions over the past millennium reveal agreement on major climatic episodes. In their fourth assessment report (AR4), IPCC Working Group I specifically addressed temperatures during the MWP and concluded from the large spread of values between individual records that there was an increased heterogeneity in temperature during the MWP (Jansen et al., 2007; IPCC, 2007). However a re-analysis of the data displayed in the IPCC Report by Esper and Frank (2009) indicated no increased spread between long-term proxies, but they concluded that an insufficient number of widespread high-resolution climate proxies is available to draw sound conclusions on the spatial extent of warmth during the MWP. For Europe, some evidence indicates an intensive MWP, although above-average temperatures were likely not evenly spread (Hughes and Diaz, 1994; Goosse et al., 2006). In contrast to temperature, little is known about the hydroclimatic conditions during the MWP, in particular for Europe. Proxy evidence from around the world indicates that the MWP hydroclimate was distinct from that of today and differently pronounced around the globe. While North America, southern South America and costal East Africa appeared to be dry, the Medieval hydroclimate was wet in northern South America, the Sahel

6.1 Introduction

region and southern Africa (Seager et al., 2007, and references therein). However, records of Medieval hydroclimate in Europe are few, of dubious quality, and often contradictory. While two speleothem records from northwest Scotland (Proctor et al., 2002) and Israel (Bar-Matthews et al., 1998) implied a wet MWP, records of floods in the Netherlands (Tol and Langen, 2000) and the Iberian peninsula (Benito et al., 2003) suggested rather dry conditions.

As high-elevation sites are known for their particular sensitivity to climatic change (Hughes and Diaz, 1994), the European Alps became target of recent advances in paleoclimate research. Systematic retrieval of records from historic buildings in the southwestern Swiss Alps (Büntgen et al., 2006a) enabled existing tree-ring width and maximum latewood density chronologies to be extended back in time, resulting in two summer temperature reconstructions that exceed the last millennium (Büntgen et al., 2005, 2006b). In addition to these annually resolved tree-ring archives, further proxies with lower temporal resolution were explored. The varved sediments of lake Silvaplana revealed autumn-temperature sensitive diatoms (Blass et al., 2007), chironomids reflecting July-temperature at near annual resolution (Larocque et al., 2009), and various mineral composition as indicators of autumn, summer and annual precipitation as well as summer temperature (Trachsel et al., 2008). However, precipitation is spatially much more heterogenous than temperature especially in complex topography (Brunetti et al., 2006). Therefore, more moisture-sensitive proxies are needed to address long-term variations in central European climate with sufficient confidence.

The climate system is mainly driven by the sun. Despite the sun's importance as driving force, little is known to what extent solar variability contributes to past and present climate change (Beer et al., 2000). For meanwhile more than three decades satellite-based measurements show that solar irradiation fluctuates in phase with solar activity and its 11-year sunspot cycle (e.g., Hoyt and Schatten, 1998). Although these irradiance variations are too small to have significant impacts on the climate, there are indications that long-term solar variability could be much larger. There is growing evidence that periods of low solar activity coincide with advances in glaciers (e.g., Denton and Karlén, 1973), changes in lake levels (e.g., Magny, 1993), and sudden changes of climatic conditions (e.g., van Geel and Renssen, 1998) and a coupling between pre-1850 temperatures and solar-activity is indicated (Crowley, 2000). Regarding a causal link between sun and climate, alterations in solar output may be accompanied by changes in the cloudiness caused by variations in the intensity of galactic cosmic rays in the atmosphere (Carslaw et al., 2002). This would help to explain why relatively small changes in solar output can produce large changes in the Earth's climate. Separating solar from non-solar induced climate changes is therefore highly desirable (e.g., Beer et al., 2000; Crowley, 2000).

Stable carbon (δ^{13}C) and oxygen (δ^{18}O) isotopes in tree rings have been used successfully to gain knowledge about the climate of the last centuries in a variety of studies (e.g., Libby et al., 1976; Burk and Stuiver, 1981; Lipp et al., 1991; Saurer et al., 1995; Anderson et al., 1998; see also McCarroll and Loader, 2004, and references therein). Stable isotope fractionation (e.g., Craig and Gordon, 1965; Farquhar et al., 1989; Ehleringer et al., 1993; Roden et al., 2000) does not reflect the net tree-growth, but integrates physical condition and tree responses. Unlike the classical dendroclimatological variables tree-ring width and maximum latewood density, stable isotope ratios are therefore not only determined by tree-growth limiting climate factors and can record additional climate information (e.g., McCarroll and Pawellek, 2001). This has been impressively demonstrated for the high-elevation Lötschental site in the Swiss Alps. While tree-ring width and maximum latewood density are clearly restricted to summer temperature (Büntgen et al., 2005, 2006b), carbon isotopes are highly summer temperature and moisture sensitive and oxygen isotopes additionally reveal a striking interconnection to summer sunshine duration (Treydte et al., 2001; Kress et al., 2010).

In this study, we address millennial-scale moisture and sunshine duration variability for the Swiss Alps, revealed by a carbon and oxygen isotope chronology from European larch (*Larix decidua* Mill.). To obtain the full 1,200 years, living trees grown at tree-line sites in the Lötschental were extended by samples obtained from historical buildings in the Lötschental and the Simplon region. We thus present the longest available tree-ring δ^{13}C and δ^{18}O records in central Europe, which are highly sensitive to central European climate changes.

6.2 Selection and processing of material

To build a millennial length climate reconstruction from tree-ring stable isotopes, the tree species and field sites had to be chosen considering three aspects: (1) the tree species is sufficiently long-living; (2) it is growing in a climatically sensitive environment; and (3) the region provides additionally historic material from the same species

6.2 Selection and processing of material

to extend the chronology beyond the life-span of the species. Based on previous studies (Büntgen et al., 2005, 2006b,a) European larch (*Larix decidua* Mill.) from the two high elevation sites Lötschental and Simplon region (compare Fig. 6.1) was chosen. The Simplon region was in particular included to improve coverage of the most distant period.

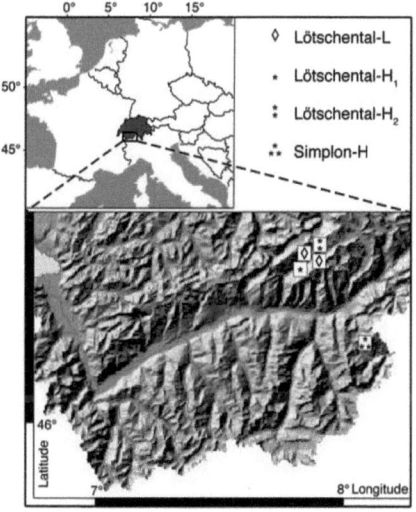

Figure 6.1: The sampling sites Lötschental (LOE) and Simplon (SIM) including living trees (L) and historical buildings (H).

Chapter 6 1200 years of climate history from stable carbon and oxygen isotopes

6.2.1 The Lötschental

The alpine valley Lötschental is situated in the Valais, southwestern Switzerland. Its upper part belongs to the "Jungfrau-Aletsch-Bietschhorn" region in the Bernese Alps, which is protected as a UNESCO natural world heritage. It is the largest, and aside from Leukerbad the only inhabited side valley on the northern side of the dry inner-alpine Rhône river. From a climatic point of view it is situated in a transition zone between the moist regime of the northern Alps and the continental inner- alpine character of the central Valais. On a distance of 28 km it preserves an altitudinal gradient of > 2500 m a.s.l. stretching from the valley exit in Steg/Gampel (634 m a.s.l.) to the Lötschenlücke (3,178 m a.s.l.) and is surrounded by several peaks > 3500 m a.s.l. with the Bietschhorn (3,939 m a.s.l.) being the highest. Because of its altitude, the vegetation is dominated by the subalpine belt of spruce-larch forests (*Larici-Picetum*; Ellenberg, 1996), which is above ∼1,900 m a.s.l. gradually replaced by larch-stone pine associations (*Larici-Pineatum cembrae*; Ellenberg, 1996), which are forming the upper tree line at ∼2,200 m a.s.l. today.

The Lötschental was likely first settled during the Roman period (Anneler, 1917), but it remained remote and difficult to access until the beginning of the 20th century, when the Lötschbergbahn connected it to the Swiss railway system. The oldest remnant forests can be found above the villages, were they have been preserved as protection against avalanches (so-called "Bannwald"). Overall, the dominant tree species is European larch (*Larix decidua* Mill.) reaching a maximum age of up to 700 years in the Lötschental (Büntgen et al., 2005). With their long and straight stems and their resistance to weathering, they provide an ideal construction timber (Schweingruber, 2001; Büntgen et al., 2004). As traditional farming was dominant during most of the continuous settlement history many buildings are preserved. While most residential houses are located near the valley floor, temporarily inhabited Alp-huts can be found up to an altitude of 2,100 m a.s.l. (Büntgen et al., 2006a).

The old-aged living larch trees, together with wood preserved in historical buildings has made the Lötschental a prominent location for dendrochronological studies, such as millennial temperature reconstructions (Büntgen et al., 2005, 2006b), dendroecological network analyses (Frank and Esper, 2005), reconstructions of larch budmoth activity (Esper et al., 2007; Kress et al., 2009a, see Chapter 4) and, more recently, the assessment of intra-annual tree growth along elevational transects (Moser et al., 2010) as well as intra-annual oxygen isotopic variation in the atmosphere-soil-plant systems (Boda et al., subm).

6.2.2 The Simplon region

The Simplon region lies roughly within 25 km of linear distance to the Lötschental. It is stretching across the southern main crest of the western Alps from Brig (671 m a.s.l.) in the Upper Valais to northern Italy traversing the Simplon-Pass at an altitude of 2,005 m a.s.l.. The climate is influenced by the dry inner-alpine Rhône valley with its continental character, the southern Alps main crest and an insubrian influence from the southern side of the Alps. The Simplon region is surrounded by peaks up to 3,553 m a.s.l (Monte Leone) in the East and up to 4,023 m a.s.l. (Weissmies) in the West. The northern slope comprises oak-beech forests (*Querco-Fagetum*; Ellenberg, 1996) at low elevations, spruce-larch forests (*Larici-Piceetum*; Ellenberg, 1996) at mid elevations and larch-stone pine associations towards the treeline (*Larici-Pineatum cembrae*; Ellenberg, 1996). The latter can also be found at higher elevations of the southern slope, while its lower elevations are dominated by fir-beech forests (*Abieti-Fagetum*; Ellenberg, 1996). The upper tree line is presently situated between 1,900 and 2,200 m a.s.l. (Müller, 2005).

The Simplon pass is one of the lowest passes of the Alps and has presumably been used for a long time. Possibly, it was already known in the Neolithic (Sauter, 1950; Welten, 1982). During the Roman period, it was enlarged either as a military or trade road and was frequently used for long distance trade between the 12th and 15th century. The settlement "Simplon" was first mentioned in historical records in AD 1240 (Müller, 2005, and references therein). The Simplon region is thus providing wood preserved in historical buildings from periods before a continuous settlement has taken place in the Lötschental. Some of these ancient buildings have been used by Büntgen et al. (2005) and Büntgen et al. (2006b) to extend their reconstructions back in time.

6.2.3 Sampling design

Larch samples were collected in four cohorts from different locations (Fig. 6.1) and represent a so-called "composite chronology", a compilation from living trees and historical timber. The very recent period is covered by living trees from two tree-line sites

Chapter 6 1200 years of climate history from stable carbon and oxygen isotopes

in the Lötschental (NNW- and SSW-facing slope; LOE-L, Fig. 6.2), followed by two periods covered by historic material from two different buildings in the Lötschental (LOE-H_1, Fig. 6.3 and LOE-H_2, Fig 6.4) and the earliest period consisting of historic material from one building in Simplon Dorf (SIM-H, Fig. 6.5). Each period is overlapping with the subsequent by at least 50 years. Details of the locations, site characteristics and covered periods can be found in Table 6.1. For all samples, tree-ring width was measured and cross-dated following standard procedures (Stokes and Smiley, 1968; Holmes, 1983) to date each tree ring to its year of formation. For each cohort five dominant trees were chosen for isotope analysis (except for LOE-H_2, where only four samples provided sufficiently well preserved tree rings), a number proven to be satisfactory for isotope analysis (McCarroll and Loader, 2004; Treydte et al., 2007).

Table 6.1: Overview of the different cohorts of trees within the chronology and their origin. Sampling and sample preparation was split between the Paul Scherrer Institut (PSI) and the University of Bern (Bern).

Location	Cohort (abbr.)	Wood type	Altitude (m asl)	Period covered	No. of trees	In charge of
Lötschental; tree-line sites (south- and north-facing)	LOE-L	living	~2100	AD 1650–2004	5	PSI
Lötschental; Gemeindestadel Blatten	LOE-H_1	historic	1540	AD 1278–1700[1]	5	Bern
Lötschental; Stallscheune Eisten-Gryn	LOE-H_2	historic	1600	AD 1180–1325[2]	4	PSI
Simplon; Stall Dorsaz-Guntern Simplon Dorf	SIM-H	historic	1480	AD 800–1183[3]	5	PSI

[1] One tree is extending back to AD 1256

[2] AD 1304-1325 is only covered by three trees; two trees are extending back to AD 1150

[3] One tree is extending forward to AD 1198

Figure 6.2: Sampling site of the living material: tree-line sites in the Lötschental (LOE-L).

Figure 6.3: Sampling site of historic material: "Gemeindestadel" in Blatten, Lötschental (LOE-H_1).

Figure 6.4: Sampling site of historic material: "Stallscheune" in Eisten-Gryn, Lötschental (LOE-H_2).

Figure 6.5: Sampling site of historic material: "Stall Dorsaz/Guntern" in Simplon Dorf, Simplon (SIM-H).

6.2.4 Isotope analysis

Each core was split year-by-year using a scalpel. Earlywood was not separated from latewood, as isotope analysis of 50 sub-samples has shown no significant difference between the two tree-ring compounds (Kress et al., 2009b, see Chapter 3). Besides the calibration period (AD 1900–2004), in which each tree was analyzed separately, all cores from one cohort were pooled prior to analysis for each annual ring, but retaining single-tree measurements every 10th year ("10-year split pool approach"). This breakup of the pool in regular intervals enables the continuous control of signal strength within the different trees. The pooling of tree rings of the same year from cores from the same site proved successful for climate analysis, retaining the annual signal but substantially reducing the workload of sample preparation (Treydte et al., 2007). To avoid an influence of a potential juvenile effect (an increase of $\delta^{13}C$ values in young trees; McCarroll and Loader, 2005, and references therein), care was taken to cut-off the first 50 years of each core. However, this was not done for the cohort of LOE-H_1 to ensure sufficient overlap between LOE-H_1 and LOE-H_2. Alpha-cellulose was extracted from all samples following standard procedures (Boettger et al., 2007) adapted for larch samples, homogenized by sonification and freeze-dried for 24 h (Kress et al., 2009a).

The cohorts LOE-L, LOE-H_2 and SIM-H were measured at the Paul Scherrer Institut (PSI) with the following setup: carbon isotopic ratios were determined with a reproducibility of 0.1‰ after combustion to CO_2 at 1025 °C in an elemental analyzer (EA-1110; Carlo Erba Thermoquest, Milan, Italy) coupled to an isotope ratio mass spectrometer (Delta S or Delta Plus XL; Thermo Finnigan Mat, Bremen, Germany). Oxygen isotopic ratios were assigned after pyrolysis to CO at 1080 °C using a continuous flow method connected via a variable open-split interface Conflo III (Thermo Finnigan Mat, Bremen, Germany) with a reproducibility of 0.3‰ (Saurer and Siegwolf, 2004).

The cohort LOE-H_1 was measured at the University of Bern with a setup where carbon and oxygen isotopic ratios were determined simultaneously by online high temperature reduction (above 1450 °C). Oxygen and carbon isotopic ratios were hereby directly analyzed after pyrolysis to CO in a thermochemical elemental analyzer, which is coupled to an isotope ratio mass spectrometer (Delta Plus XL) via a Conflo II device (all Thermo Finnigan Mat, Bremen, Germany) (Knoller et al., 2005; Leuenberger and Filot, 2007). The reproducibility is 0.5‰ for oxygen and 0.3‰ for carbon isotopic ratios.

6.3 Results

All isotopic ratios are expressed as delta notation relative to an international standard (VPDB for δ^{13}C and VSMOW for δ^{18}O). The carbon isotope series were corrected for the atmospheric decline in δ^{13}C due to fossil fuel burning since the beginning of the industrialization (Leuenberger, 2007).

6.2.5 Statistical analysis

The climate-isotope relationships were determined using the Swiss part of the HISTALP-dataset (Auer et al., 2007) in a detailed sensitivity study with the LOE-L material (Kress et al., 2010, see Chapter 5). To further investigate signal persistence in both isotope series, we conducted high and low pass filtering using cubic smoothing spline with 50% frequency response cut-off at 10 years and 30 years (Cook and Peters, 1981). In addition, spectral analysis was performed using the multi-taper method (MTM, 2 year resolution, 3 tapers; Mann and Lees, 1996) of the spectra software (Ghil et al., 2002). Finally, to assess the longer-term trends of the climate-isotope signal in the context of existing reconstructions, we compared the hundred most positive and negative isotope values of the entire period with the corresponding events in the maximum latewood density (MXD) series by Büntgen et al. (2006b). This MXD record consists of 180 larch samples, mainly collected in the Lötschental and Simplon region, and provides a strong summer temperature signal.

6.3 Results

6.3.1 The raw carbon and oxygen chronologies

For both isotope series, the composite picture of measured values reveal large offsets between the different cohorts (Fig. 6.6). This rather inconsistent pattern between cohorts is somewhat stronger expressed in carbon than oxygen isotopes. While for carbon the living material (LOE-L) provides the lowest values compared to all three cohorts from historical buildings (LOE-H$_1$, LOE-H$_2$, SIM-H), oxygen does not repeat this picture as LOE-H$_1$ possesses the lowest mean (Fig 6.6; Tab. 6.2). Although originating from a slightly different site, the isotope ratios of SIM-H are within the range of all three LOE-cohorts. At the same time both isotope series of LOE-H$_1$,

Chapter 6 1200 years of climate history from stable carbon and oxygen isotopes

which were measured with a different setup (see Chapter 6.2.4) do not significantly differ from all other cohorts. In addition to the offset between different cohorts, there is also a change in variance. This is again slightly more obvious in the carbon than oxygen isotope series, but in both isotope series LOE-L and LOE-H_2 reveal a higher variance as LOE-H_1 and SIM-H (compare Tab. 6.2). This was not expected, as LOE-H_1 and LOE-H_2 both represent historical buildings in the Lötschental that are not far apart.

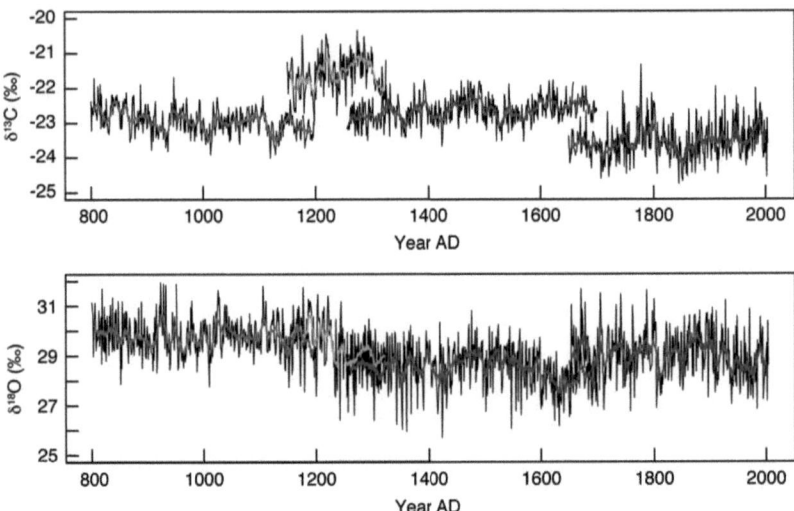

Figure 6.6: Annually resolved carbon (top) and oxygen (bottom) chronologies for AD 800-2004, composed of the different tree-cohorts from past to present: SIM-H (green), LOE-H_2 (orange), LOE-H_1 (blue) and LOE- L (red). Bold lines represent 11-year moving windows (centered). All carbon isotope values have been corrected for the decline in δ^{13}C of atmospheric CO_2.

6.3 Results

Table 6.2: Statistical numbers of carbon and oxygen isotopic composition in the different cohorts of the chronology measured at the Paul Scherrer Insitut (PSI) and the University of Bern (Bern).

Cohort	Period covered	Mean δ^{13}C (‰)	Variance δ^{13}C (σ^2)	Mean δ^{18}O (‰)	Variance δ^{18}O (σ^2)	Measured by
LOE-L	AD 1650–2004	-23.54	0.27	29.02	0.95	PSI
LOE-H$_1$	AD 1278–1700[1]	-22.65	0.14	28.49	0.74	Bern
LOE-H$_2$	AD 1180–1325[2]	-21.59	0.27	29.41	0.99	PSI
SIM-H	AD 800–1183[3]	-22.94	0.15	29.79	0.53	PSI

[1] One tree is extending back to AD 1256

[2] AD 1304-1325 is only covered by three trees; two trees are extending back to AD 1150

[3] One tree is extending forward to AD 1198

To exclude differences in isotope values between two cohorts caused by different measurement setups (see Chapter 6.2.4), the first overlap period (op1, AD 1650–1700) was measured with both systems from the identical material (LOE-L, Fig. 6.7). Although seven value are missing in the measurements performed by Bern (AD 1650, 1659, 1660, 1670, 1680, 1690, 1700), both isotope curves progress in the same way as those measured by PSI and reveal the preservation of the variance as well as a very similar overall mean (-23.30‰ for carbon and 29.24‰ for oxygen compared to -23.54‰ for carbon and 29.02‰ for oxygen). This is in agreement with a comparison which has been performed with standardized wood material (Tab. 6.3). In a small case study, isotopes were measured with both setups, from alpha-cellulose of standardized wood material extracted in both laboratories. Regardless of which laboratory extracted cellulose and performed measurements, the isotope measurements were virtually identical. The standard deviation between three measurements of 0.05 for carbon and 0.09 for oxygen is clearly below the range at which measurements can be repoduced (compare Chapter 6.2.4).

Chapter 6 1200 years of climate history from stable carbon and oxygen isotopes

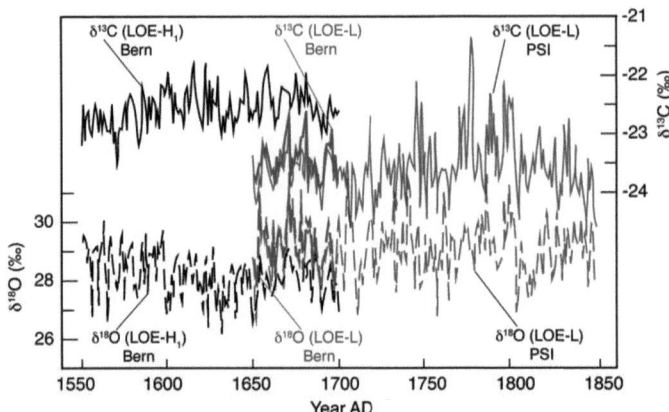

Figure 6.7: Carbon and oxygen isotopes in the first overlap period (op1, AD 1650–1700). LOE-L was measured at PSI and LOE-H$_1$ in Bern. The entire op1 from LOE-L was additionally measured in Bern (red lines) but without AD 1650, 1659, 1660, 1670, 1680, 1690 and 1700.

Table 6.3: Inter-laboratory comparison of alpha-cellulose of standardized wood material extracted and measured at Paul Scherrer Institut (PSI) and the University of Bern (Bern).

Inter-laboratory comparison standard wood material (pine standard Reynolds)	δ^{13}C (‰)	δ^{18}O (‰)
Cellulose extraction PSI[1] and measurements PSI	-26.39	27.83
Cellulose extraction Bern[2] and measurements Bern	-26.47	27.67
Cellulose extraction Bern and measurements PSI	-26.38	27.69
Standard deviation of measurements	0.05	0.09

[1] PSI: Paul Scherrer Institut; carbon measurements are performed by combustion
[2] Bern: University of Bern; carbon measurements are performed by pyrolysis

6.3.2 Linking the cohorts: the standardized chronologies

As no clear pattern in the offset between isotope values of different cohorts is visible and the offset cannot be explained by different measurement setups, it seems risky to adjust the older cohorts to the level of the recent period. If all cohorts were adjusted to the level of the living trees, an overall decreasing trend in carbon and an overall increasing trend in oxygen would show up. As such strong trends are very likely not reflecting any long term climate trends, they over-emphasize an artificial signal. Thus, we standardised each cohort over its entire length to the same mean (=0) and standard deviation (=1), before linking the overlap periods by simply taking the average of both cohorts (Fig. 6.8).

While most overlap periods display a reasonably good agreement (see Fig. 6.8), the carbon isotpes in overlap period 2 (AD 1256–1325) diverge before AD 1300. This may be caused by the occurence of a "juvenile effect" (McCarroll and Loader, 2005, and references therein), as the cores of LOE-H_1 were analysed until the very beginning (see Chapter 6.2.4). Therefore, we decided to reduce this overlap period 2 to AD 1300–1325 so that the period before AD 1300 is represented only by LOE-H_2 instead of the average of both cohorts. By standardizing the isotope chronologies, the variance was set to 1, such that events within individual cohorts can be compared directly.

6.3.3 The climate signal(s) in the standardized chronologies

When a chronology is standardized before linking together its different parts, there is a potential risk to lose lower frequency information extending the length of each "segment" (Cook et al., 1995). To consider the preservation of properties in the lower frequency domain, the standardized isotope series were analyzed using the multi-taper method over the 800-2004 period (see Chapter 6.2.5). As expected, spectral power was preserved in frequencies above ∼0.5 cycles/year (data not shown), but there were also significant spectral power below ∼0.05 cycles/year (Fig. 6.9). In this low frequency domain, carbon isotopes reveal three highly significant (99% confidence level) peaks at ∼60, ∼114 and ∼186 years, while one highly significant peak at ∼603 years was evident from the oxygen isotope series. These findings are rather surprising as the shortest segment within the chronology (LOE-H_2) contains 176 years. In addition, the oxygen isotopes possess another highly significant peak at 11.5 years, which may

Chapter 6 1200 years of climate history from stable carbon and oxygen isotopes

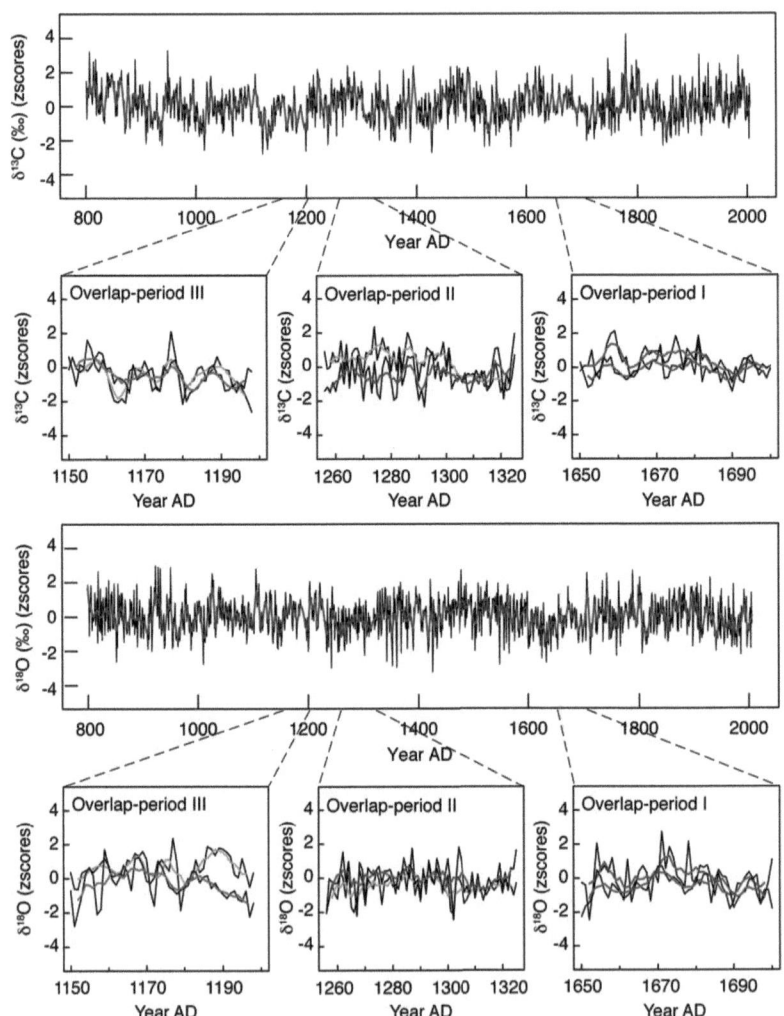

Figure 6.8: Carbon (top) and oxygen (bottom) chronologies and their overlap periods after standardization of each segment to the same mean (=0) and standard deviation (=1). Bold lines represent 10-year low pass filtered series based on cubic smoothing splines with 50% frequency response cut-off at 10 years. In the long chronologies the overlap periods are represented as a mean between two segments whereas in the close-ups both segments of each overlap period are shown. The color code in the close-ups corresponds to the cohorts in Figure 6.6.

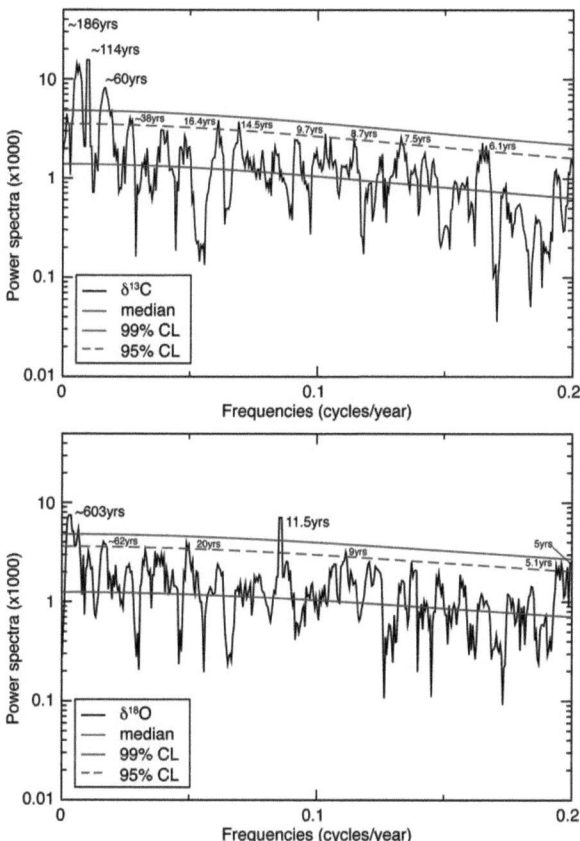

Figure 6.9: Multi-taper method (MTM) spectra using two-year resolution/ three tapers of $\delta^{13}C$ (top) and $\delta^{18}O$ (bottom) for AD 800–2004. Both isotope series were standardized to the same mean (=0) and standarddeviation (=1) before MTM was performed. Large letters indicate peaks (in years=yrs) above the 99% confidence level (CL), smaller letters mark peaks above the 95% CL.

Chapter 6 1200 years of climate history from stable carbon and oxygen isotopes

reflect the solar cycle as the average duration of the sunspot cycle is about 11 years (e.g., Hoyt and Schatten, 1998). Further peaks at a 95% significance level referring to decadal (up to centennial) trends are present in both isotope series (see Fig. 6.9). While both isotope series demonstrate decadal to centennial trends, a preservation of signal beyond the minimum segment length is rather unlikely in the carbon but possibly given in the oxygen series.

30-year low pass filtered series (see Chapter 6.2.5) revealed different patterns for both standardized isotope series (Fig. 6.10). The most visible difference was the range of amplitude, which is considerably smaller for oxygen than for carbon isotopes. While the 20th century reveals positive anomalies in the carbon series, the oxygen series shows negative ones. The opposite longer-term trends in the two isotope series are continuing back in time and are, besides the 20th century, particularly pronounced during the 16th and 17th centuries. In contrast, the isotopes reveal a coherent picture faintly expressed in the 9th century and strongly indicated during the 14th, 15th and 18th century. The overall pattern reveals a higher number of peaks in the oxygen series but stronger and more distinct peaks in the carbon series.

If we combine the information obtained from this standardized isotope series, with results of a sensitivity study performed for LOE-L (Kress et al., 2010, see Chapter 5) and an additional maximum latewood density (MXD) record of the Lötschental (Büntgen et al., 2006b), we can address climate signals and trends (Fig. 6.11). Hereby, it is noteworthy that the three proxies are sensitive to slightly different seasons. While the carbon and the oxygen series revealed a very narrow window restricted to July and August (Kress et al., 2010, see Chapter 5) , the MXD series shows the best climate response to a prolonged period from June to September (Büntgen et al., 2006b). In Figure 6.11 the hundred most positive and negative events (= annual values) in both isotope and the MXD series were added up according to their century of occurrence.

According to Büntgen et al. (2006b) and Kress et al. (2010), the MXD-series can be interpreted as temperature, the carbon series as drought and the oxygen series as a sunshine signal. The resulting picture illustrates the most heterogenous distribution in the MXD and the most homogenous distribution in the oxygen isotope series. Most extreme events (n = 34) can be found in the MXD series during the 20th century, reflecting the recent warming trend. The isotope series reveal the 9th century as extremely dry (19 events) and sunny (15 events) and the 12 th century was, according to

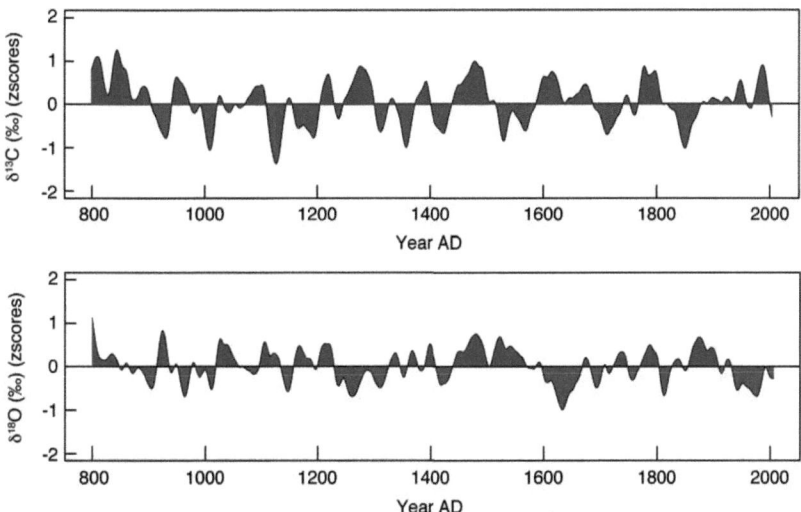

Figure 6.10: Carbon (top) and oxygen (bottom) chronologies after applying a 30-year low pass filter based on cubic smoothing splines with a 50% frequency response cut-off at 30 years. While periods with above-average positive values are indicated in red, those with below-average negative values are colored in blue.

the carbon isotopes, the wettest century (20 events). The century that appears most unusual in all three parameters is the 17th century with least sunshine, no wet and most cold extreme events, which may coincide with the the "Little Ice Age" (LIA; Overpeck et al., 1997). In contrast, the 13th century, although also rather cloudy, appears warm and dry, which may be connected to the late "Medieval Warm Period" (MWP; Hughes and Diaz, 1994).

When combining the information from Figure 6.10 and 6.11, the last 1200 years can be interpreted from past to present as the following. In the 9th century, it was rather moderately warm, but extremely sunny and dry. This period was followed by a much warmer 10th century with average amounts of sunshine and a quite wet appearance. While there was not much of a change in sunshine duration, it clearly

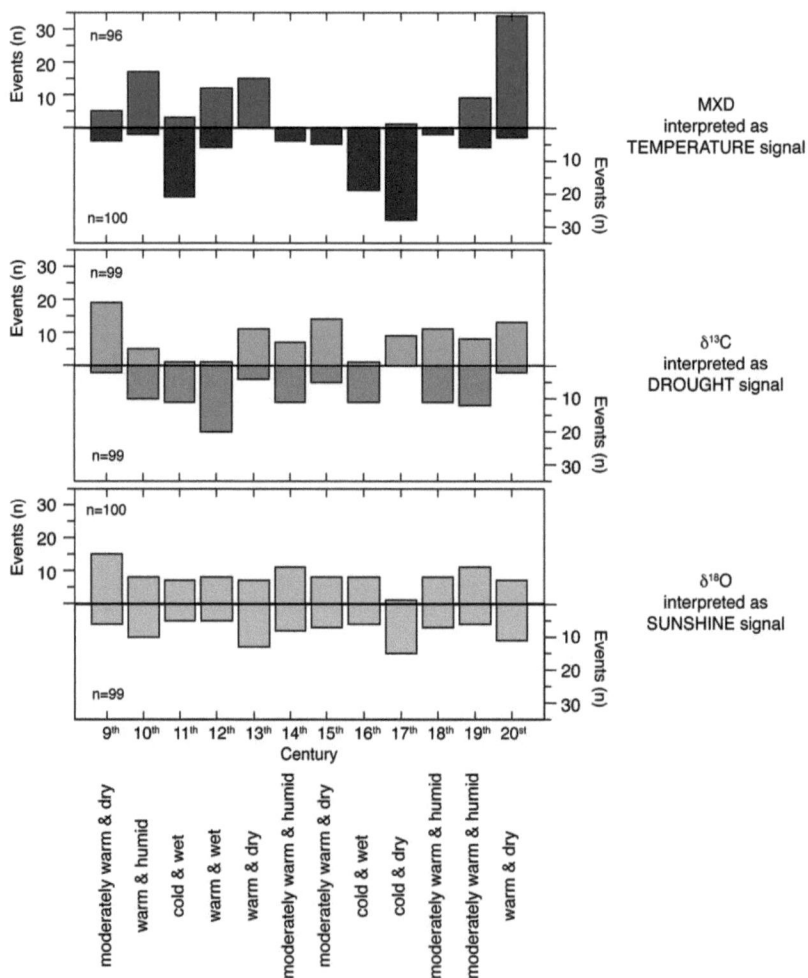

Figure 6.11: Extreme events in maximum latewood density (MXD) (top), δ^{13}C (middle) and δ^{18}O (bottom) for AD 800–2004. For each parameter, the 100 most positive and negative values (n = 100) were chosen and added up for each century. MXD is interpreted as temperature, δ^{13}C as drought and δ^{18}O as sunshine signal. Extreme events in the 21st century are not shown (n < 100 in some cases).

became cold and wetter during the 11th century. While slowly getting warmer during the 12th century, the wet character further increased. This roughly three-century long wet period was disrupted in the 13th century, which was warm and dry, while being relatively cloudy. During the 14th to 16th century it was getting colder while being sunny and wet except for the dry second half of the 15th century. The climate of the 17th century is different from the entire period as being extremely cold and at the same time dry and cloudy. From the 18th century onwards, the trend in temperature is rising, while being indifferent in sunshine and rather wet during the first half of the 18th, the second half of the 18th revealed sunny and dry conditions. The entire 19th century appeared to be sunny and wet while slowly getting warmer. Compared to the entire period, the 20th century is extremely warm, rather dry and at the same time relatively cloudy.

The comparison of the two isotope series (Fig 6.10) reveals periods of agreement and mismatch in longer-term trends. If the additional temperature information of the MXD series (Fig. 6.11) is drawn on this comparison, a temperature-dependent pattern is revealed. The longer-term trends in the isotope series agree as long as the temperature stays moderate. This is the case during the 9th, 14th, 15th, 18th and 19th centuries. As soon as extremes in temperature occur (either positive or negative ones), the isotope series show a diverging behavior in their longer-term trends, as shown in the 16th, 17th and 20th centuries.

6.4 Discussion

The tree-ring isotope records presented here are the longest currently available chronologies for western Europe, and they are moisture sensitive (carbon) and reveal sunshine variability (oxygen). As most previous climate reconstructions have been focussing on temperature (Jansen et al., 2007), the new information provided here is an important step towards a comprehensive understanding of long-term changes in Europe's climate.

6.4.1 Climate signals

Although we cannot distinguish major climate episodes such as the MWP, LIA, and recent warmth from the stable isotopes alone, we provide additional decadal to centen-

Chapter 6 1200 years of climate history from stable carbon and oxygen isotopes

nial information of the coinciding hydrological cycle. From the isotope perspective, the most outstanding centuries were the 9th being extremely dry and sunny, the very wet 12th followed by the cloudy but dry 13th, the dry and sunny 15th, the dry and cloudy 17th and the dry 20th century. If we compare this with length fluctuations of the Great Aletsch glacier, large retreats are reported around AD 800 and for the 20th century, and maximum extensions in the 14th and 17th century (Haeberli and Holzhauser, 2003). This strengthens our hypothesis of a dry and sunny pre-900 period and 20th century not just being very warm but at the same time rather dry. The 17th century appears to be rather dry and cloudy. This may at first glance interfere with an advance in glacier length, but as the 17th century appeared to be cold (e.g., Büntgen et al., 2006b), we assume that the obviously dry conditions were caused by rather short summers with precipitation falling partially as snow. In addition, there are seasonal discrepancies. As the isotope series are linked to a very narrow window during the vegetation period (Kress et al., 2010, see Chapter 5), they do not capture the full annual signal, which has to be taken into account when interpreting phenomena that are linked to the entire hydrological cycle. For western Europe, dry summers during most of the 13th century have also been reported by Hughes and Diaz (1994), consistent with a phase of alpine glacier retreat, as noted by Grove and Switsur (1994). This period was followed by cold springs and humid winters coupled with rather wet summers (Hughes and Diaz, 1994), which are also evident in the carbon isotope series, resulting in glacier retreats that are evident e.g., in records from the Great Aletsch glacier (Haeberli and Holzhauser, 2003). Despite the difference in seasonality, alpine glacier advances and retreats seem to support the overall picture of wet and dry summer conditions given by the carbon isotope series.

If we compare the MWP with the recent warming trend from the perspective of the isotope records, the 20th century is unusual with respect to temperature, but it is not with respect to drought occurences and the amount of sunshine. The MWP, however, appeared to be very wet for a major period ($\sim$1000–1200 AD). This is contrary to assumptions of rather dry conditions in Europe (e.g., Tol and Langen, 2000; Benito et al., 2003) but is in agreement with other studies, which found predominantly wet conditions in Europe within that period (Bar-Matthews et al., 1998; Proctor et al., 2002). A very recent study revealed a persistent positive Northern Atlantic Oscillation (NAO) signal during the MWP (Trouet et al., 2009). This signal was likely accompanied by increased precipitation over the Atlantic and northwestern Europe and decreased precipitation in

6.4 Discussion

south-central Europe and in the Mediterranean. Although the NAO contains mainly a winter signal, this strong and consistent NAO may have also affected summer hydroclimate in Europe. Therefore, westerlies may have predominantly influenced the climate in the summer, resulting in higher precipitation amounts in (north-)western Europe, which would agree with our findings.

To date, tree-ring oxygen isotope series have played a minor role in the interpretation of past climate conditions. Besides showing high correlations with summer sunshine duration and very high inter-series correlations indicating a strong driving force (Kress et al., 2010, see Chapter 5), the oxygen isotope series revealed a highly significant peak of 11.5 years in its frequency spectra (Fig. 6.9). This 11.5-year periodicity may reflect solar cycle alterations as the average duration of the sunspot cycle is about 11 years (Schove, 1983; Hoyt and Schatten, 1998). Another well-known cycle related to disturbances of the sun (magnetic polarity reversal of the sunspots) is the so-called Hale cycle (Wilson, 1988). With a periodicity of 22-23 years it is exactly twice as long as the 11.5 year cyclicity found in the oxygen record. In addition, the great solar minima of the last millennium with multidecadal duration, i.e., Wolf (centered $\sim$1305), Spörer (centered $\sim$1470), Maunder (centered $\sim$1680), Dalton (centered $\sim$1810) and recent (since 1960) (Stuiver and Braziunas, 1989; Usoskin et al., 2007) are, except for Spörer, clearly reflected in the oxygen record.

The connection between tree-ring width and solar activity has long been investigated, but no significant and consistent relationship was identified (LaMarche and Fritts, 1972). However, Saurer et al. (2000) reported significant interdecadal variations in oxygen isotope series with a periodicity of $\sim$24 years in a 158-year record of oxygen isotopes from *Abies alba* grown in the Swiss Jura. Although they considered sunspot cycles as a possible reason for periodic variations in climate parameters, they concluded that these $\sim$24 years periodicities are likely caused by fluctuations in the large-scale atmospheric circulation over Europe and the North Atlantic Ocean. Indeed, the air mass origin is to a certain extent reflected in the oxygen isotopes, as $\delta^{18}O$ records faithfully the isotopic composition of meteoric precipitation (Saurer et al., 1997; Rebetez et al., 2003). The latter, in turn, is highly variable according to the air mass properties (e.g., temperature) and is therefore a strong indicator for its provenance (Dansgaard, 1964; Rozanski et al., 1993). $\delta^{18}O$ may therefore reflect changes in the synoptic setting, which can be expressed in sunshine duration (see also Kress et al., 2010, Chapter 5). At the same time, sunshine duration is strongly linked to the amount of shortwave radiation entering the atmosphere. This incoming shortwave radiation has a major impact on leaf-surface temperature and is affecting the process of leaf water enrichment (e.g., Barbour et al., 2004) much more strongly than ambient air temperatures. Therefore an additional sunshine signal is likely to be be imprinted upon $\delta^{18}O$. This sunshine-$\delta^{18}O$ relationship may become of particular interest in the ongoing debate to what extent vulcanic eruptions (and their accompanying greenhouse gas emission and aerosol production) are affecting climate changes in the longer term (e.g., Stothers, 1984; McCromick et al., 1995; Briffa et al., 1998; Trigo et al., 2009). The oxygen isotopes seem to prove a relationship between solar activity and climate and may become of great value in the context of separating natural from anthropogenic forcing (Beer et al., 2000; Crowley, 2000).

6.4.2 Strengths and remaining limitations of the standardized isotope series

Chronologies obtained from stable isotopes have the overall advantage that the correlations between trees from one site are usually very strong, so that the signal-to-noise ratio is high and relatively few trees are required to provide a representative average (Robertson et al., 1997; McCarroll and Loader, 2004; Gagen et al., 2007; Treydte et al., 2007). Our carbon and oxygen series are based on five trees for most of the time and on 4 trees throughout the entire 1200 years. Unlike many chronologies from the classical physical tree-ring parameters (tree-ring width and maximum latewood density), our isotope chronologies do not significantly vary in sample size. If, for example, sample size was constantly decreasing (e.g., in tree-ring width studies on living trees) the mean chronology's signal-to-noise ratio would decrease as well, which can result in an increase of artificial variance back in time (Frank et al., 2007; Lee et al., 2008). We can therefore anticipate a constant signal-to-noise ratio in both isotope chronologies, although the variance between the raw data of different cohorts is not identical (see Tab. 6.2).

Within all cohorts, we attempted to sample trees within the same age class to avoid an uneven distribution of juvenile and mature growth levels as climate-sensitivity of physical tree-ring properties can be age-dependent (Esper et al., 2008). Although the full age of the trunks from historical buildings could not always determined (e.g.,

6.4 Discussion

because of the cut-off of outermost parts to shape the beam), a minimum age of 300+ years is certain. Furthermore, a recent climate-sensitivity comparison of stable carbon and oxygen isotopes from older (450–550 year) and younger (150–250 year) trees (Kress et al., 2010, see Chapter 5) has shown that both age classes preserve the climate signal equally well. As the final data series does not include any evidence from the first 50 years of tree growth, any juvenile effects from depleted carbon values can be excluded (e.g., McCarroll and Pawellek, 2001; Raffalli-Delerce et al., 2004; Gagen et al., 2007). Any uncertainties linked to different age structures within the cohorts or to a possible juvenile effect can therefore be neglected.

The greatest advantage of tree-ring stable isotope series is, possibly, that they do not contain any long-term age related trends and thus do not require statistical detrending (McCarroll and Loader, 2004; Gagen et al., 2007), the probably most problematic constraint of tree-ring width, which has become known as the so-called "segment length curse" (Cook et al., 1995). This decline in ring-width caused by the increasing circumference of the trunk can be statistically removed by regression-based techniques or alternative methods (e.g., regional curve standardization), but low-frequency climate signals occurring over the life-span of the tree (the segment length) will be affected (Cook et al., 1995; Esper et al., 2002). Gagen et al. (2007) have shown that no such "segment length curse" seem to be present in carbon isotopes. As both of the isotope series demonstrate large offsets between the different cohorts of living and historical but also between different cohorts of historical material (in particular for carbon; see Fig. 6.6 and Tab 6.2), this may only be true as long as the dataset contains simply living trees and they were standardized over a common period.

The off-sets between different cohorts can not be explained by a different measurement setup (compare Chapter 6.2.4, Fig. 6.7 and Tab. 6.3), and they do not reveal a constant trend back in time (Fig. 6.6). Ideally, these offsets would be eliminated by standardizing the values for each cohort using differences from the mean over a common period (McCarroll and Pawellek, 1998), but no such common period is available. Applying the so called "mean-adjustment" for the overlap period of two cohorts from present to past revealed implausible decreasing trends in the carbon and increasing trends in the oxygen series (data not shown). We therefore standardized each cohort over its entire segment length. The resulting chronologies are very likely to retain all of the high to mid frequency variability (Fig. 6.9) expressed in annual and decadal to centennial trends, but may not preserve the entire low-frequency variability

Chapter 6 1200 years of climate history from stable carbon and oxygen isotopes

(multi-centenntial trends) as the standardizing procedure does not preserve the original variance.

To overcome this limitation, new detrending methods may need to be developed. One suggested solution is the so called "trig-point method" (M. Gagen pers. comm., 2009). In this method the "true" cohort-mean is determined by numerous measurements (20+) of specimens at the beginning and the end of each cohort. Between these two values a linear fit is adjusted to detrend the entire cohort. After detrending each cohort according to their "true" means, the cohorts are linked by mean-adjustment. Nevertheless, this procedure requires a high sample coverage throughout the investigated period, which is not realistic when dealing with material from historical buildings. Further limitations are the assumption that the mean of a cohort is changing in a linear manner from its beginning to its end. This may be indeed an oversimplification as a non-linear variability is most likely and errors caused by this procedure may be many times higher than the true low-frequency amplitude.

Lastly, it is noteworthy that the climatically relevant content in both stable isotope series is limited to a rather short window within the vegetation period (Kress et al., 2010, see Chapter 5). Although this July-August summer signal is highly significant and of great value to assess summer conditions, it is clearly not representative for all-season climate variability. Thus, conclusions on changing climate conditions on a year-round basis cannot be drawn.

6.5 Conclusions

The major climatic periods MWP, LIA and the recent warming trend as well as their transition intervals can be further differentiated in wet/dry and sunny/cloudy interdecadal variations by the interpretation of the carbon and oxygen isotope records. Despite reliably covering decadal to centennial variations, multi-centennial variations are likely to diminish or even disappear by linking the different cohorts via standardization. To preserve the entire spectra of climate frequencies in stable isotope series from living and historic material new detrending approaches would need to be applied.

Nevertheless, the carbon and oxygen archives contain important information, in particular about past climate variability. While carbon is faithfully recording the July-August interplay of temperature and moisture conditions, oxygen is strongly linked to summer sunshine duration. According to the carbon isotopes, the MWP was rather wet in contrast to the predominantly dry recent warming. Together with the prevailing warm conditions, the MWP may have provided favorable conditions for agricultural production and therefore human subsistence in central Europe. The sunshine signal in the oxygen isotopes is indicating a periodicity at ~11 years, which is likely to reflect the solar cycle and may provide further evidence for a relationship between solar activity and climate forcing. Together the isotope chronologies may help to constrain climate predictions and might therefore contribute to plan for, adapt to and mitigate impacts of future climate change.

Acknowledgements

This work was funded by the EU project FP6-2004-GLOBAL-017008-2 (MILLENNIUM). Thanks to A. Verstege, D. Nievergelt and M. Schmidhalter for support in the field and Dendro-LAB and to M. Tröndle, L. Läubli and N. Bircher for assistance with sample preparation.

References

Anderson, W. T., Bernasconi, S. M., McKenzie, J. A., and Saurer, M. (1998). Oxygen and carbon isotopic record of climatic variability in tree ring cellulose (*Picea abies*): an example from central Switzerland (1913-1995). *Journal of Geophysical Research*, 103(D24):31625–31636; doi:10.1029/1998JD200040.

Anneler, H. (1917). *Lötschen das ist: Landes- und Volkskunde des Lötschentals*. Akademische Buchhandlung von Max Drechsel, Bern.

Auer, I., Böhm, R., Jurkovic, A., Lipa, W., Orlik, A., Potzmann, R., Schoner, W., Ungersbock, M., Matulla, C., Briffa, K., Jones, P., Efthymiadis, D., Brunetti, M., Nanni, T., Maugeri, M., Mercalli, L., Mestre, O., Moisselin, J. M., Begert, M., Muller-Westermeier, G., Kveton, V., Bochnicek, O., Stastny, P., Lapin, M., Szalai, S., Szentimrey, T., Cegnar, T., Dolinar, M., Gajic-Capka, M., Zaninovic, K., Majstorovic, Z., and Nieplova, E. (2007). HISTALP - historical instrumental climatological surface time series of the Greater Alpine Region. *International Journal of Climatology*, 27(1):17–46; doi:10.1002/joe.1377.

Bar-Matthews, M., Ayalon, A., and Kaufmann, A. (1998). Middle to late Holocene (6,500 yr. period) paleoclimate in the eastern Mediterranean region from the stable isotopic composition of speleotherms from Sorec cave, Israel. In Issar, A. S. and Brown, N., editors, *Water, Environment and Scociety in Times of Climatic Change*, pages 203–214. Kluwer Academic Publishers, Dordrecht.

Barbour, M. M., Roden, J. S., Farquhar, G. D., and Ehleringer, J. R. (2004). Expressing leaf water and cellulose oxygen isotope ratios as enrichment above source water reveals evidence of a Péclet effect. *Oecologia*, 138(3):426–435; doi:10.1007/s00442-003-1449-3.

Beer, J., Mende, W., and Stellmacher, R. (2000). The role of the sun in climate forcing. *Quaternary Science Reviews*, 19(1-5):403–415; doi:10.1016/S0277-3791(99)00072-4.

Benito, G., Díez-Herrero, A., and Fernández de Villalta, M. (2003). Magnitude and frequency of flooding in the Tagus basin (central Spain) over the last millennium. *Climatic Change*, 58(1):171–192; doi:10.1023/A:1023417102053.

Blass, A., Bigler, C., Grosjean, M., and Sturm, M. (2007). Decadal-scale autumn temperature reconstruction back to AD 1580 inferred from the varved sediments of Lake Silvaplana (southeastern Swiss Alps). *Quaternary Reserarch*, 68(2):184–195; doi:10.1016/j.yqres.2007.05.004.

Boda, S. Y., Treydte, K. S., Fonti, P., Gessler, A., Graf-Pannatier, E., Saurer, M., Siegwolf, R. T. W., and Werner, W. (subm.). Intra-seasonal pathway of oxygen isotopes from soil to wood in the Loetschental (Swiss Alps). *Plant, Cell and Environment*.

Boettger, T., Haupt, M., Knoller, K., Weise, S. M., Waterhouse, J. S., Rinne, K. T., Loader, N. J., Sonninen, E., Jungner, H., Masson-Delmotte, V., Stievenard, M., Guillemin, M. T., Pierre, M., Pazdur, A., Leuenberger, M., Filot, M., Saurer, M., Reynolds, C. E., Helle, G., and Schleser, G. H. (2007). Wood cellulose preparation methods and mass spectrometric analyses of $\delta^{13}C$, $\delta^{18}O$ and nonexchangeable δ^2H values in cellulose, sugar, and starch: an interlaboratory comparison. *Analytical Chemistry*, 79(12):4603–4612; doi:10.1021/ac0700023.

Briffa, K. R., Jones, P. D., Schweingruber, F. H., and Osborn, T. J. (1998). Influence of volcanic eruptions on Northern Hemisphere summer temperature over the past 600 years. *Nature*, 393(6684):450–455; doi:10.1038/30943.

Brunetti, M., Maugeri, M., Nanni, T., Auer, I., Bohm, R., and Schoner, W. (2006). Precipitation variability and changes in the greater Alpine region over the 1800-2003 period. *Journal of Geophysical Research*, 111(D11):D11107; doi:10.1029/2005JD006674.

Büntgen, U., Brazdil, R., Frank, D., and Esper, J. (2009). Three centuries of Slovakian drought dynamics. *Climate Dynamics*, pages in press; doi:10.1007/s00382-009-0563-2.

Büntgen, U., Esper, J., Frank, D. C., Nicolussi, K., and Schmidhalter, M. (2005). A 1052-year tree-ring proxy for Alpine summer temperatures. *Climate Dynamics*, 25(2-3):141–153; doi:10.1007/s00382-005-0028-1.

Büntgen, U., Esper, J., Schmidhalter, M., Frank, D. C., Treydte, K., Neuwirth, B., and Winiger, M. (2004). Using recent and historical larch wood to build a 1300-year Valais chronology. In Gärtner, H., Esper, J., and Schleser, G. H., editors, *TRACE - Tree Rings in Archaeology, Climatology and Ecology*, volume 2, pages 85–92.

Büntgen, U., Frank, D. C., Bellwald, I., Kalbermatten, H., Freund, H., Schmidhalter, M., Bellwald, W., Neuwirth, B., and Esper, J. (2006a). 700 years of settlement and building history in the Lötschental/Valais. *Erdkunde*, 60:96–112; doi:10.3112/erdkunde.2006.02.02.

Büntgen, U., Frank, D. C., Niervergelt, D., and Esper, J. (2006b). Summer temperature variations in the European Alps, AD 755-2004. *Journal of Climate*, 19(21):5606–5623; doi:10.1175/JCLI3917.1.

Burk, R. and Stuiver, M. (1981). Oxygen isotope ratios in trees reflect mean annual temperature and humidity. *Science*, 211:1417–1419; doi:10.1126/science.211.4489.1417.

Carslaw, K. S., Harrison, R. G., and Kirkby, J. (2002). Cosmic rays, clouds, and climate. *Science*, 298(5599):1732–1737; doi:10.1126/science.1076964.

Cook, E. R., Briffa, K. R., Meko, D. M., Graybill, D. A., and Funkhouser, G. (1995). The segment length curse in long tree-ring chronology development for paleoclimatic studies. *The Holocene*, 5(2):229–237; doi:10.1177/095968369500500211.

Cook, E. R. and Peters, K. (1981). The smoothing spline: a new approach to standardizing forest interior tree-ring width series for dendroclimatic studies. *Tree Ring Bulletin*, 41:45–53.

Craig, H. and Gordon, L. (1965). Deuterium and oxygen^{18} variations in the ocean and marine atmospheres. In Tongoirgi, E., editor, *Proceedings of a Conference on Stable Isotopes in Oceanographic Studies and Palaeotemperatures*, pages 9–130, Pisa, Italy. Lischi and Figli.

Crowley, T. J. (2000). Causes of climate change over the past 1000 years. *Science*, 289(5477):270–277; doi:10.1126/science.289.5477.270.

Dansgaard, W. (1964). Stable isoptopes in precipitation. *Tellus*, 16:436–468.

Denton, G. H. and Karlén, W. (1973). Holocene climatic variations–their pattern and possible cause. *Quaternary Research*, 3(2):155–174; doi:10.1016/0033-5894(73)90040-9.

Ehleringer, J. R., Hall, A. E., and Farquar, G. D. (1993). *Stable Isotopes and Plant Carbon-Water Relations*. Academic Press, New York.

Ellenberg, H. (1996). *Vegetation Mitteleuropas mit den Alpen in ökologischer, dynamischer und historischer Sicht*. Ulmer Verlag, Stuttgart, 5th edition.

Esper, J., Büntgen, U., Frank, D. C., Niervergelt, D., and Liebhold, A. (2007). 1200 years of regular outbreaks in alpine insects. *Proceedings of the Royal Society B*, 274:671–679; doi:10.1098/rspb.2006.0191.

Esper, J., Cook, E. R., and Schweingruber, F. H. (2002). Low-frequency signals in long tree-ring chronologies for reconstructing past temperature variability. *Science*, 295(5563):2250–2253; doi10.1126/science.1066208.

Esper, J. and Frank, D. (2009). The IPCC on a heterogeneous Medieval Warm Period. *Climatic Change*, 94(3):267–273; doi:10.1007/s10584-008-9492-z.

Esper, J., Niederer, R., Bebi, P., and Frank, D. (2008). Climate signal age effects-evidence from young and old trees in the Swiss Engadin. *Forest Ecology and Management*, 255(11):3783–3789; doi:10.1016/j.foreco.2008.03.015.

Esper, J., Wilson, R. J. S., Frank, D. C., Moberg, A., Wanner, H., and Luterbacher, J. (2005). Climate: past ranges and future changes. *Quaternary Science Reviews*, 24(20-21):2164–2166; doi:10.1016/j.quascirev.2005.07.001.

6.5 Conclusions

Farquhar, G. D., Ehleringer, J. R., and Hubick, K. T. (1989). Carbon isotope discrimination and photosynthesis. *Annual Review of Plant Physiology and Plant Molecular Biology*, 40:503–537; doi:10.1146/annurev.pp.40.060189.002443.

Frank, D. and Esper, J. (2005). Characterisazion and climate response patterns of a high-elevation multi-species tree-ring network in the European Alps. *Dendrochronologia*, 22:107–121; doi:10.1016/j.dendro.2005.02.004.

Frank, D., Esper, J., and Cook, E. R. (2007). Adjustment for proxy number and coherence in a large-scale temperature reconstruction. *Geophysical Research Letters*, 34(16):L16709; doi:10.1029/2007GL030571.

Gagen, M., McCarroll, D., Loader, N. J., Robertson, L., Jalkanen, R., and Anchukaitis, K. J. (2007). Exorcising the 'segment length curse': summer temperature reconstruction since AD 1640 using non-detrended stable carbon isotope ratios from pine trees in northern Finland. *The Holocene*, 17(4):435–446; doi:10.1177/0959683607077012.

Ghil, M., Allen, M. R., Dettinger, M. D., Ide, K., Kondrashov, D., Mann, M. E., Robertson, A. W., Saunders, A., Tian, Y., Varadi, F., and Yiou, P. (2002). Advanced spectral methods for climatic time series. *Reviews of Geophysics*, 40(1):doi:10.1029/2000RG000092.

Goosse, H., Arzel, O., Luterbacher, J., Mann, M. E., Renssen, H., Riedwyl, N., Timmermann, A., Xoplaki, E., and Wanner, H. (2006). The origin of the European "Medieval Warm Period". *Climate of the Past*, 2(2):99–113.

Grove, J. M. and Switsur, R. (1994). Glacial geological evidence for the Medieval Warm Period. *Climatic Change*, 26(2-3):143–169; doi:10.1007/BF01092411.

Haeberli, W. and Holzhauser, H. (2003). Alpine glacier mass changes during the past two millennia. *Pages News*, 11(1):13–15.

Holmes, R. L. (1983). Computer-assisted quality control in tree-ring dating and measurements. *Tree-Ring Bulletin*, 43:69–78.

Hoyt, D. V. and Schatten, K. H. (1998). Group sunspot numbers: a new solar activity reconstruction. *Solar Physics*, 179(1):189–219; doi:10.1023/A:1005007527816.

Hughes, M. K. and Diaz, H. F. (1994). Was there a "Medieval Warm Period", and if so, where and when? *Climatic Change*, 26(2-3):109–142; doi:10.1007/BF01092410.

Huntington, T. G. (2006). Evidence for intensification of the global water cycle: review and synthesis. *Journal of Hydrology*, 319(1-4):83–95; doi:10.1016/j.jhydrol.2005.07.003.

IPCC (2007). *Climate Change 2007: the Physical Science Basis. Contribution of Working Group I to the Fourth Assessment Report of the Intergovernmental Panel on Climate Change*. Cambridge University Press, Cambridge, United Kingdom and New York, NY, USA.

Jansen, E., Overpeck, J., Briffa, K. R., Duplessy, J.-C., Joos, F., Masson-Delmotte, V., Olago, D., Otto-Bliesner, B., Peltier, W. R., Rahmstorf, S., Ramesh, R., Raynaud, D., Rind, D., Solomina, O., Villalba, R., and Zhang, D. (2007). Palaeoclimate. In Solomon, S., Qin, D., Manning, M., Chen, Z., Marquis, M., Averyt, K., Tignor, M., and Mille, H., editors, *Climate Change 2007: The Physical Science Basis. Contribution of Working Group I to the Fourth Assessment Report of the Intergovernmental Panel on Climate Change*, pages 433–497. Cambridge University Press, Cambridge, United Kingdom and New York, NY, USA.

Knoller, K., Boettger, T., Weise, S. M., and Gehre, M. (2005). Carbon isotope analyses of cellulose using two different on-line techniques (elemental analysis and high-temperature pyrolysis) - a comparison. *Rapid Communications in Mass Spectrometry*, 19(3):343–348; doi:10.1002/rcm.1793.

Kress, A., Saurer, M., Büntgen, U., Treydte, K., Bugmann, H., and Siegwolf, R. T. W. (2009a). Summer temperature dependency of larch budmoth outbreaks revealed by Alpine tree-ring isotope chronologies. *Oecologia*, 160(2):353–365; doi:10.1007/s00442-009-1290-4.

Kress, A., Saurer, M., Siegwolf, R. T. W., Esper, J., Frank, D. C., and Bugmann, H. (2010). A 350-year drought reconstruction from an Alpine tree-ring stable isotope chronology. *Global Biogeochemical Cycles*, 24:GB2011; doi:10.1029/2009GB003613.

Kress, A., Young, G. H. F., Saurer, M., Loader, N. J., Siegwolf, R. T. W., and McCarroll, D. (2009b). Stable isotope coherence in the earlywood and latewood of tree-line conifers. *Chemical Geology*, 268(1-2):52–57; doi:10.1016/j.chemgeo.2009.07.008.

Kundzewicz, Z., Mata, L., Arnell, N., Döll, P., Kabat, P., Jiminéz, B., Miller, K., Oki, T., Sen, Z., and Shiklomanov, I. (2007). Freshwater resources and their management. In Parry, M., Canziani, O., Palutikof, J., van der Linden, P. J., and Hanson, C.,

6.5 Conclusions

editors, *Climate Change 2007: Impacts, Adaptation and Vulnerybility. Contribution of Working Group II to the Fourth Assessment Report of the Intergovernmental Panel on Climate Change*, pages 173–210. Cambridge University Press, Cambridge, United Kingdom and New York, NY, USA.

LaMarche, V. J. and Fritts, H. C. (1972). Tree-rings and sunspot numbers. *Tree-Ring Bulletin*, 32:19–33.

Lamb, H. (1965). The early medieval warm epoch and its sequel. *Palaeogeography, Palaeoclimatology Palaeoecology*, 1:13–37; doi:10.1016/0031-0182(65)90004-0.

Larocque, I., Grosjean, M., Heiri, O., Bigler, C., and Blass, A. (2009). Comparison between chironomid-inferred July temperatures and meteorological data AD 1850-2001 from varved Lake Silvaplana, Switzerland. *Journal of Paleolimnology*, 41(2):329–342; doi:10.1007/s10933-008-9228-0.

Lee, T., Zwiers, F., and Tsao, M. (2008). Evaluation of proxy-based millennial reconstruction methods. *Climate Dynamics*, 31(2):263–281; doi:10.1007/s00382-007-0351-9.

Leuenberger, M. (2007). To what extent can ice core data contribute to the understanding of plant ecological developments of the past? In Dawson, T. and Siegwolf, R., editors, *Stable Isotopes as Indicators of Ecological Change*, pages 211–233. Elsevier Academic Press, London.

Leuenberger, M. and Filot, M. (2007). Temperature dependencies of high-temperature reduction on conversion products and their isotopic signatures. *Rapid Communications in Mass Spectrometry*, 21:1587–1598; doi:10.1002/rcm.2998.

Libby, L. M., Pandolfi, L. J., Payton, P. H., Marshall III, J., Becker, B., and Giertz-Sienbenlist, V. (1976). Isotopic tree thermometers. *Nature*, 261(284-288; doi:10.1038/261284a0).

Lipp, J., Trimborn, P., Fritz, P., Moser, H., Becker, B., and Frenzel, B. (1991). Stable isotopes in tree ring cellulose and climatic change. *Tellus B*, 43B:322–330; doi:10.1034/j.1600-0889.1991.t01-2-00005.x.

Magny, M. (1993). Solar influences on Holocene climatic changes illustrated by correlations between past lake-level fluctuations and the atmospheric ^{14}C record. *Quaternary Reserarch*, 40(1):1–9; doi:10.1006/qres.1993.1050.

Mann, M. E. and Lees, J. (1996). Robust estimation of background noise and signal detection in climatic time series. *Climatic Change*, 33:409–445; doi:10.1007/BF00142586.

McCarroll, D. and Loader, N. J. (2004). Stable isotopes in tree rings. *Quaternary Science Reviews*, 23(7-8):771–801; doi:10.1016/j.quascirev.2003.06.017.

McCarroll, D. and Loader, N. J. (2005). *Isotopes in Tree Rings*, volume 10 of *Developments in Paleoenvironmental Research Series*, chapter 2, pages 67–116. Springer.

McCarroll, D. and Pawellek, F. (1998). Stable carbon isotope ratios of latewood cellulose in *Pinus sylvestris* from northern Finland: variability and signal-strength. *The Holocene*, 8(6):675–684; doi:10.1191/095968398675987498.

McCarroll, D. and Pawellek, F. (2001). Stable carbon isotope ratios of *Pinus sylvestris* form northern Finland and the potential for extracting a climate signal from long Fennoscandian chronologies. *The Holocene*, 11(5):517–526; doi:10.1191/095968301680223477.

McCromick, M. P., Thomason, L. W., and Trepte, C. R. (1995). Atmospheric effects of the Mt Pinatubo eruption. *Nature*, 373(6513):399–404; doi:10.1038/373399a0.

Moser, L., Fonti, P., Büntgen, U., Esper, J., Luterbacher, J., Franzen, J., and Frank, D. (2010). Timing and duration of European larch growing season along an altitudinal gradient in the Swiss Alps. *Tree Physiology*, 30(2):225–233; doi:10.1093/treephys/tpp108.

Müller, H.-N. (2005). *Landschaftsgeschichte Simplon (Walliser Alpen Schweiz): Gletscher-, Vegetations- und Klimaentwicklung seit der Eiszeit*, volume 17 of *Karlsruher Schriften zur Geographie und Geoökologie*. Karlsruhe.

Overpeck, J., Hughen, K., Hardy, D., Bradley, R., Case, R., Douglas, M., Finney, B., Gajewski, K., Jacoby, G., Jennings, A., Lamoureux, S., Lasca, A., MacDonald, G., Moore, J., Retelle, M., Smith, S., Wolfe, A., and Zielinski, G. (1997). Arctic environmental change of the last four centuries. *Science*, 278(5341):1251–1256; doi:10.1126/science.278.5341.1251.

Proctor, C. J., Baker, A., and Barnes, W. L. (2002). A three thousand year record of North Atlantic climate. *Climate Dynamics*, 19(5):449–454; doi:10.1007/s00382-002-0236-x.

Raffalli-Delerce, G., Masson-Delmotte, V., Dupouey, J. L., Stievenard, M., Breda, N., and Moisselin, J. M. (2004). Reconstruction of summer droughts using tree-ring cellulose isotopes: a calibration study with living oaks from Brittany (western France). *Tellus B*, 56(2):160–174; doi:10.1111/j.1600–0889.2004.00086.x.

Rebetez, M., Saurer, M., and Cherubini, P. (2003). To what extent can oxygen isotopes in tree rings and precipitation be used to reconstruct past atmospheric temperature? A case study. *Climatic Change*, 61(1-2):237–248; doi:10.1023/A:1026369000246.

Robertson, I., Switsur, V. R., Carter, A. H. C., Barker, A. C., Waterhouse, J. S., Briffa, K. R., and Jones, P. D. (1997). Signal strength and climate relationships in $^{13}C/^{12}C$ ratios of tree ring cellulose from oak in east England. *Journal of Geophysical Research*, 102(D16):19507–19516; doi:10.1029/97JD01226.

Roden, J. S., Lin, G., and Ehleringer, J. R. (2000). A mechanistic model for interpretation of hydrogen and oxygen isotope ratios in tree-ring cellulose. *Geochimica et Cosmochimica Acta*, 64(1):21–35; doi:10.1016/S0016–7037(99)00195–7.

Rozanski, K., Arguas-Arguas, L., and Gonfiantini, R. (1993). Isotopic patterns in modern global precipitation. In Swart, P., editor, *Climate Change in Continental Isotopic Records*, volume 78 of *Geophysical Monograph*, pages 1–36. American Geophysical Union, Washington, DC.

Saurer, M., Borella, S., Schweingruber, F. H., and Siegwolf, R. (1997). Stable carbon isotopes in tree rings of beech: climatic versus site-related influences. *Trees - Structure and Function*, 11(5):291–297; doi:10.1007/s004680050087.

Saurer, M., Cherubini, P., and Siegwolf, R. (2000). Oxygen isotopes in tree rings of abies alba: The climatic significance of interdecadal variations. *Journal of Geophysical Research*, 105(D10):12,461–12,470; doi:10.1029/2000JD900160.

Saurer, M., Siegenthaler, U., and Schweingruber, F. H. (1995). The climate-carbon isotope relationship in tree-rings and the significance of site conditions. *Tellus B*, 47(3):320–330; doi:10.1034/j.1600–0889.47.issue3.4.x.

Saurer, M. and Siegwolf, R. (2004). Pyrolysis techniques for oxygen isotope analysis of cellulose. In *Handbook of Stable Isotope Analytical Techniques*, volume 1, pages 497–508; doi:10.1016/B978–044451114–0/50025–9. Elsevier, New York.

Sauter, M.-R. (1950). Préhistoire du valais des origines aux temps mérovingiens. In *Vallesia*, volume V, pages 1–165.

Schove, D. (1983). *Sunspot cycles*, volume 68 of *Benchmark Papers in Geology*. Stroudsburg, PA, Hutchinson Ross Publishing Co.

Schweingruber, F. H. (2001). *Dendroökologische Holzanatomie. Anatomische Grundlagen der Dendrochronologie*. Eidgenössische Forschungsanstalt WSL.

Seager, R., Graham, N., Herweijer, C., Gordon, A. L., Kushnir, Y., and Cook, E. (2007). Blueprints for medieval hydroclimate. *Quaternary Science Reviews*, 26(19-21):2322–2336; doi:10.1016/j.quascirev.2007.04.020.

Stokes, M. A. and Smiley, T. L. (1968). *An introduction to tree-ring dating*. (reprinted 1996). University of Arizona Press, Tucson, US, Chicago.

Stothers, R. B. (1984). The great Tambora eruption in 1815 and its aftermath. *Science*, 224(4654):1191–1198; doi:0.1126/science.224.4654.1191.

Stuiver, M. and Braziunas, T. F. (1989). Atmospheric ^{14}C and century-scale solar oscillations. *Nature*, 338(6214):405–408; doi:10.1038/338405a0.

Tol, R. S. J. and Langen, A. (2000). A concise history of Dutch river floods. *Climatic Change*, 46(3):357–369; doi:10.1023/A:1005655412478.

Trachsel, M., Eggenberger, U., Grosjean, M., Blass, A., and Sturm, M. (2008). Mineralogy-based quantitative precipitation and temperature reconstructions from annually laminated lake sediments (Swiss Alps) since AD 1580. *Geophysical Research Letters*, 35(13):L13707; doi:10.1029/2008GL034121.

Treydte, K., Frank, D., Esper, J., Andreu, L., Bednarz, Z., Berninger, F., Boettger, T., D'Alessandro, C. M., Etien, N., Filot, M., Grabner, M., Guillemin, M. T., Gutierrez, E., Haupt, M., Helle, G., Hilasvuori, E., Jungner, H., Kalela-Brundin, M., Krapiec, M., Leuenberger, M., Loader, N. J., Masson-Delmotte, V., Pazdur, A., Pawelczyk, S., Pierre, M., Planells, O., Pukiene, R., Reynolds-Henne, C. E., Rinne, K. T., Saracino, A., Saurer, M., Sonninen, E., Stievenard, M., Switsur, V. R., Szczepanek, M., Szychowska-Krapiec, E., Todaro, L., Waterhouse, J. S., Weigl, M., and Schleser, G. H. (2007). Signal strength and climate calibration of a European tree-ring isotope network. *Geophysical Research Letters*, 34(24):L24302; doi:10.1029/2007GL031106.

6.5 Conclusions

Treydte, K. S., Schleser, G. H., Schweingruber, F. H., and Winiger, M. (2001). The climatic significance of $\delta^{13}C$ in subalpine spruces (Lötschental, Swiss Alps). *Tellus B*, 53(5):593–611; doi:10.1034/j.1600-0889.2001.530505.x.

Trigo, R., Vaquero, J., Alcoforado, M.-J., Barriendos, M., Taborda, J., Garcìa-Herrera, R., and Luterbacher, J. (2009). Iberia in 1816, the year without a summer. *International Journal of Climatology*, 29(1):99–115; doi:10.1002/joc.1693.

Trouet, V., Esper, J., Graham, N. E., Baker, A., Scourse, J. D., and Frank, D. C. (2009). Persistent positive north atlantic oscillation mode dominated the medieval climate anomaly. *Science*, 324(5923):78–80; doi:10.1126/science.1166349.

Usoskin, I., Solanki, S. K., and Kovaltsov, G. A. (2007). Grand minima and maxima of solar activity: new observational constraints. *A&A*, 471(1):301–309; doi:10.1051/0004-6361:20077704.

van Geel, B. and Renssen, H. (1998). Abrupt climate change around 2,650 BP in North-West Europe: evidence for climatic teleconnections and a tentative explanation. In Issar, A. S. and Brown, N., editors, *Water, Environment and Scociety in Times of Climatic Change*, volume 31, pages 21–41. Kluwer Academic Publishers, Utrecht.

Welten, M. (1982). *Vegetationsgeschichtliche Untersuchungen in den westlichen Schweizer Alpen: Bern–Wallis*, volume 95 of *Denkschriften der Schweizerischen Naturforschenden Gesellschaft*. Basel.

Wilson, R. M. (1988). Bimodality and the Hale cycle. *Solar Physics*, 117(2):269–278: doi:10.1007/BF00147248.

7

Conclusions and outlook

In this PhD thesis, two tree-ring stable isotope chronologies from European larch (*Larix decidua* Mill.) were developed. This chapter will discuss these currently longest available carbon and oxygen isotope series for central Europe in the context of:

(1) methodological challenges,

(2) stable isotopes in European larch as climate archive,

(3) the biological "bias" and its ecological relevance, and

(4) future perspectives.

Chapter 7 Conclusions and outlook

Methodological challenges

European larch is a widespread tree species throughout the higher elevations in the European Alps. With a longevity of 850+ years, its widespread utilization as timber and its high temperature sensitivity (Carrer and Urbinati, 2006), it is considered to be an ideal archive for climate reconstructions. While the more traditional dendroclimatological variables tree-ring width and maximum latewood density have shown to be reliable summer temperature proxies (e.g., Büntgen et al., 2005, 2006), the suitability of tree-ring carbon and oxygen isotopes of larch as a natural archive has been constrained to date by methodological limitations.

Old-aged larch trees possess very narrow rings ($\ll 0.5$mm), making the accurate separation of rings demanding and reliable isotope measurements from the minute amount of cellulose almost impossible. In addition, larch is periodically infested by the foliage-feeding grey larch budmoth, resulting in discernable alterations of cell properties in the tree rings. Besides affecting the tree-ring, and therefore cellulose quantity, it was unclear if these outbreak events would also affect the isotopic signatures, thereby masking climate signals. To overcome these limitations and to make tree-ring isotopes in larch accessible as a climate archive, the following three challenges were addressed:

1. *Optimizing the sample preparation procedure*: As a major material loss is resulting from the conventional homogenization procedure of milling the tree rings prior to cellulose extraction (Borella et al., 1998), this step was replaced by sonification of cellulose (Chapter 4). In contrast to the milling procedure, with losses up to 30%, the losses during sonification and the subsequent freeze-drying are negligible. In addition, the procedure of cellulose extraction was optimized for small quantities of wood, resulting in a yield of up to 35% of alpha-cellulose (its content in wood amounts to about 40%; Borella et al., 1998). Furthermore, the resulting highly homogeneous cellulose enabled measurements with half the sample amount (0.2 mg for δ^{13}C and 0.5 mg for δ^{18}O) compared to the conventional procedure (Borella et al., 1998; Saurer and Siegwolf, 2004), while preserving the reproducibility of measurements of 0.1‰ for δ^{13}C and 0.3‰ for δ^{18}O.

2. *Testing for homogeneity in tree-ring earlywood and latewood*: An intercomparison of the isotopic composition in tree-ring earlywood, latewood and whole-ring

demonstrated a high common signal between earlywood and latewood and indicated reliable climate calibrations from all three wood constituents. As whole-ring cellulose may even improve the climate correlations, its use for climate reconstructions is fully justified (Chapter 3).

3. *Assessing the influence of larch budmoth outbreaks*: Several lines of evidence were provided that larch budmoth outbreaks do not affect the isotope ratios in tree rings of larch, neither during the outbreak nor in subsequent years (Chapter 4). Thus, the isotope ratios preserve the climate signal during the outbreak events. Unlike tree-ring width and maximum latewood density, which contain a reliable climate signal only after appropriate corrections for larch budmoth infestations (Esper et al., 2007), no correction needs to be applied to the stable isotope series.

In sum, the stable isotope archive in European larch can be reliably assessed at an annual resolution, as measurements even from small sample amounts can be replicated with a high degree of confidence. In addition, the available sample amount increased by using whole ring-cellulose and by the optimization of sample preparation. Furthermore, no corrections are needed for years of larch budmoth infestations, which may result in a more complete retrieval of (climatically relevant) information from this unique long-living archive.

Stable isotopes in European larch as climate archive

Paleoclimatic research is of great importance, as changes in the past may help to assess future climate variability. Despite the huge variety of climate proxies, temperature-sensitive archives are considerably over-represented and many proxies may not record other variables that are of interest to climatologists (Hughes, 2002). As precipitation arguably plays a key role for human subsistence as well as for many terrestrial ecosystems, proxies recording changes in the hydrological cycle are of great importance. However, it is currently uncertain how much the recent global change is already affecting precipitation amounts and patterns (IPCC, 2007). Such changes in the global rate and distribution of precipitation may have greater direct impact on human well-being and terrestrial ecosystems than the change in temperature itself (Kundzewicz et al.,

Chapter 7 Conclusions and outlook

2007). The interpretation of the climatic signal in both isotope records is therefore discussed in this context. Thereby, the focus is set firstly on the dominating climate signals and possible reconstructions, and secondly on remaining uncertainties.

The dominating climate signals and possible reconstructions

The drought reconstruction from carbon isotopes

The carbon isotope series has indicated a high affinity to July-August temperature and at the same time to precipitation. As a time-dependent relationship between these two variables could not be ruled out (Chapter 5), the focus was set on the reconstruction of a climate variable that can account for potential instabilities in the relationship between temperature and precipitation. A comparison of three monthly resolved self-calibrating Palmer drought severity index datasets (see Chapter 5) demonstrated very different signals for the central European Alps. As none of these datasets provided a convincing record, a simple drought index (Bigler et al., 2006), incorporating monthly precipitation and temperature data, was the method of choice for the Lötschental.

The effectiveness of very simple linear models widely used in tree-ring research (Hughes, 2002) also proved true for the carbon-drought index relationship, resulting in strong calibration and verification statistics. Instead of the commonly applied ± 2 root mean squared error (RSME) as error estimate, the mean between tree correlation was divided by the RSME to account for changes in inter-tree variability. The resulting drought reconstruction contains a moisture signal for the European Alps with regional extent, which is a small, yet crucial step towards a comprehensive understanding of past changes in Europe's hydroclimate.

The sunshine signal in the oxygen isotopes

Since the oxygen isotope series showed a stronger sensitivity to sunshine than to any other climate variable (Chapter 5), this relationship was investigated further (Chapter 6). A ~11-year periodicity in the oxygen record suggested the replication of cycles in solar activity. This finding points to an active role of the sun in past climate changes and may contribute to distinguishing between solar and non-solar induced climate changes in the past. Such a separation is highly desirable as it would enable the recent warming to be put into the context of natural vs. anthropogenic climate forcing (e.g., Beer et al., 2000).

Assessing major climate periods in the past millennium

With the help of the decadal to centennial climate signals in carbon and oxygen isotopes, major climate periods of the past millennium ("Medieval Warm Period" and "Little Ice Age") that had been reconstructed previously were complemented by a novel record of wet and dry episodes as well as radiation intensity (Chapter 6). The most outstanding episodes occurred during the Medieval Warm Period, in which roughly 200 years were predominantly wet and were only disrupted by short intervals of drier conditions. Along with high temperatures, this period is likely to have been particularly favorable for agricultural production and therefore human subsistence.

Remaining uncertainties

The nature of the proxies

An optimum paleoenvironmental record of climate over the last thousand years or more would have the following characteristics: it would be (i) continuous throughout the millennium and preferably even longer; (ii) well understood in its spatial applicability; (iii) geographically wide-spread; and it would (iv) contain a strong and well-defined climate signal of known seasonal sensitivity; (v) possess a temporal resolution of one year or better; (vi) be time-invariant in strength and nature of climate signals and (vii) be capable of recording century scale as well as inter-annual and decadal scale variations (after Hughes and Diaz, 1994; Hughes, 2002).

Our tree-ring stable isotope series clearly fulfill the criteria (i)–(v) (Chapter 5, 6), but they do not accomplish in a satisfactory manner the two last criteria (vi, vii). It could neither be proven wether the climate-isotope relationships remain time-invariant beyond the instrumental period (Chapter 5) nor wether the standardized chronologies retain multi-centennial trends (Chapter 6). The considerable offsets between isotopic values in different cohorts may furthermore indicate that the usually recommended four to five trees (McCarroll and Loader, 2004; Treydte et al., 2007) do not generate a "true" mean (Chapter 6). The "offset problem" between different cohorts may possibly be alleviated by more measurements in the overlapping period of two cohorts. Finally,

Chapter 7 Conclusions and outlook

one last point has to be mentioned: the climatically relevant response remains limited to a very narrow seasonal window (Chapter 5). Although this July-August summer signal is of great importance and with a certain extent accounts for other seasons, as different seasons are not completely independent from each other, interpretations exceeding this particular time window need to be viewed with caution.

Reconstruction methods

Many studies have attempted to reconstruct northern hemisphere mean temperatures from various climate proxies (see Chapter 1). However, their results vary considerably. This may be due to the nature of the proxies (see above), but as different statistical methods have been employed in these reconstructions, the question remains how much of this discrepancy is caused by the variation in methods (Chapter 6). Although the reliability of some of the reconstruction methods has been investigated in various studies (e.g., von Storch et al., 2004; Esper et al., 2005; Mann et al., 2005; Lee et al., 2008), no overall conclusions can be drawn so far.

The biological "bias" and its ecological relevance

Besides the strong climate signals found in both isotope series, the earlywood-latewood intercomparison and the detection of larch budmoth infestations indicated two more findings of ecological relevance.

1. The highly homogeneous signals in conifer earlywood and latewood from tree-line sites may also shed light into the carbon-cycle at tree-line sites (Chapter 3). As both investigated species (*Pinus sylvestris* L. and *Larix decidua* Mill.) grow at climatically-limited sites, the close isotopic similarity in earlywood and latewood is explained by the prevailing site conditions and internal carbon budget rather than by the tree species alone. High turnover rates and small reserve pools may thus account for such a high common signal in the wood constituents. Hence, this finding may contribute to the understanding of carbon allocation in tree-line conifers and is in line with findings in Handa et al. (2005) that larch is carbon-limited at tree-line sites.

2. The results are also highly relevant for the population dynamics of the larch budmoth. When considering a summer temperature reconstruction spanning

more than three centuries (Casty et al., 2005), it becomes evident that severe larch budmoth infestations are often related to cold summers (Chapter 4). These patterns were confirmed in data from Baltensweiler and Rubli (1999) for the Valais and Engadin regions. Thus, cool July-August temperatures promote larch budmoth outbreaks. This was rather surprising as outbreaks are expected to be triggered earlier in the vegetation period (Baltensweiler et al., 1977). However, warm summers may interfere with (i) the fixed amount of energy available for egg development, and (ii) the fine-tuned relationship between needle-maturation and larvae evolution, as the needles provide the only food source for the larvae (see Chapter 4). This trend is amplified by the recent warming and consistent with the fact that since the early 1980s no alpine-wide synchronized larch budmoth event has occurred. This absence of massive outbreaks is the longest detected within the last 1,200 years (Esper et al., 2007). It remains to be seen how the recent warming trend will influence the prominent forest disturbance phenomenon of larch budmoth infestations.

Future perspectives

Already several years ago Hughes (2002) declared that the state of the art of dendroclimatology is vibrant, with "much robust debate and innovative work". When consulting the latest IPCC report (IPCC, 2007) and a recent review on paleoclimate (Jones et al., 2009), it becomes clear that advances in dendroclimatology remain a hot topic. In particular, a more realistic assessment of reconstruction uncertainty is needed. Considering all possible sources of error, robust methods should ideally "incorporate uncertainties of individual proxies, the effect of proxy selection and uncertainties in the statistical methods" (Jones et al., 2009). However, even though being highly desirable, such methods do not yet exist as advancements have not sufficiently succeeded in integrating all demands.

Composite records (containing different proxies of various origin) may contribute to overcoming the errors that are characteristic of particular proxies. Understanding the signal in each individual proxy is thereby a prerequisite (Hughes, 2002). When comparing the proxies directly among each other, before any reconstruction method has been applied, the homogeneity of climate signals in different proxies can be addressed

Chapter 7 Conclusions and outlook

and errors associated with different reconstruction methods (e.g., Lee et al., 2008) are negligible. By using trees as physical archives and combining stable isotopes with other measurements in a multiproxy approach, it should be possible to increase the precision of paleoclimate estimates, extend the range of paleoenvironmental signals that can be extracted and also test some of the assumptions inherent in paleoenvironmental reconstructions. First approaches within the Millennium project of composite spatial field correlations between uni-proxy records and a climate variable present in all records indicates a strengthening of the explained variance (Millennium, 2009). At the same time, some proxies may contain a more direct signal than others (e.g., as shown in this thesis, the carbon isotopes provide a coupling between temperature and precipitation). These differences are crucial to characterize as they may influence the ability of different proxies in scoring the target climate variable. To account for these differences, future comparisons with model simulations should contain spatially-located data rather than regional averages and, in addition, should rather directly involve the proxy data than the reconstructed variables (Edwards et al., 2007). Within the Millennium project, the European Alps are one of the areas of prime interest, which is reflected in a large array of proxies with different temporal resolution. The Alpine archives are ranging from various climate indicators in lake sediments of Lake Silvaplana in the Engadin (e.g., Blass et al., 2007; Trachsel et al., 2008; Larocque et al., 2009, over a glacio-chemical record derived from an ice core from Colle Gnifetti (Valais) (Sigl et al., prep) to annually resolved tree ring records (Büntgen et al., 2005, 2006, and this study, cf. Chapters 5, 6). Although assessing slightly different climate signals and seasons, they all demonstrate a high climate sensitivity, and further inter-comparisons are highly desirable to reveal a comprehensive picture of past Alpine climate. Still, it remains exciting, which picture of Europe's climate over the past Millennium will be revealed by the entire proxy-ensemble of the Millennium project.

Up to present, anthropogenic effects on tree-ring isotopes are an underrepresented field in tree-ring research. Although the influence of increasing nutrition and ozone on tree-ring $\delta^{13}C$ has already been indicated over a decade ago (Saurer et al., 1995), little progress has been made. At the same time, dendroclimatologists more often found an increasing discrepancy between isotope records (mainly carbon) and climate variables for the later 20th century. Indeed, there is some evidence that this results from the increasing amount of atmospheric CO_2 (McCarroll et al., 2009), but other anthropogenic influences such as nitrogen deposition cannot be excluded. Saurer et al.

(2004), for example, have shown an increased δ^{15}N in tree rings of *Picea abies* along a motorway in Switzerland, indicating an uptake of NO_x from car exhaust. As the 20th century is usually used as the calibration period and climate reconstructions are therefore constrained by the 20th century proxy-target relationship, neglecting a potential anthropogenic influence within the calibration period may result in considerable reconstruction errors. Thus, efforts for constraining anthropogenic influences on tree-ring isotope ratios are highly desirable, but they were beyond the scope of this thesis.

Overall, in this thesis two new tree-ring stable isotope chronologies were developed at an annual resolution covering a period of 1,200 years, thus representing the longest tree-ring isotope records so far available in Central Europe. The climate-isotope relationships presented in this study (Chapter 5, 6) did not show an indication of an additional anthropogenic influence in the late 20th century, as all correlations with climate variables remained stable over the entire century. This is an important discovery that strengthens the degree of confidence in the variability assessed by both isotopes. Thus, in contrast to former assumptions, tree-ring stable isotopes at the investigated tree-line site are clearly capable of capturing a climate signal different from the signal in the more traditional dendroclimatological variables tree-ring width and maximum latewood density. Despite being still quite laborious, the highly valuable information on past moisture variability and the possibility to link solar activity with climate changes justify isotope measurements on a large scale.

References

Baltensweiler, W., Benz, G., Bovey, P., and Delucchi, V. (1977). Dynamics of larch bud moth populations. *Annual Review of Entomology*, 22:79–100; doi:10.1146/annurev.en.22.010177.

Baltensweiler, W. and Rubli, D. (1999). Dispersal: an important driving force of the cycling population dynamics of the larch budmoth, *Zeiraphera diniana* Gn. In Swiss Federal Institute for Forest, S. and Landscape Research (WSL), B., editors, *Forest snow and landscape research*, volume 74, page 153. Paul Haupt, Berne, Stuttgart, Vienna.

Beer, J., Mende, W., and Stellmacher, R. (2000). The role of the sun in climate forcing. *Quaternary Science Reviews*, 19(1-5):403–415; doi:10.1016/S0277-3791(99)00072-4.

Bigler, C., Braker, O. U., Bugmann, H., Dobbertin, M., and Rigling, A. (2006). Drought as an inciting mortality factor in Scots pine stands of the Valais, Switzerland. *Ecosystems*, 9(3):330–343; doi:10.1007/s10021-005-0126-2.

Blass, A., Bigler, C., Grosjean, M., and Sturm, M. (2007). Decadal-scale autumn temperature reconstruction back to AD 1580 inferred from the varved sediments of Lake Silvaplana (southeastern Swiss Alps). *Quaternary Research*, 68(2):184–195; doi:10.1016/j.yqres.2007.05.004.

Borella, S., Leuenberger, M., Saurer, M., and Siegwolf, R. T. W. (1998). Reducing uncertainties in $\delta^{13}C$ analysis of tree rings: pooling, milling, and cellulose extraction. *Journal of Geophysical Research*, 103(D16):19,519–19,526; doi:10.1029/98JD01169.

Büntgen, U., Esper, J., Frank, D. C., Nicolussi, K., and Schmidhalter, M. (2005). A 1052-year tree-ring proxy for Alpine summer temperatures. *Climate Dynamics*, 25(2-3):141–153; doi:10.1007/s00382-005-0028-1.

Büntgen, U., Frank, D. C., Niervergelt, D., and Esper, J. (2006). Summer temperature variations in the European Alps, AD 755-2004. *Journal of Climate*, 19(21):5606–5623; doi:10.1175/JCLI3917.1.

Carrer, M. and Urbinati, C. (2006). Long-term change in the sensitivity of tree-ring growth to climate forcing in *Larix decidua*. *New Phytologist*, 170(4):861–871; doi:10.1111/j.1469-8137.2006.01703.x.

Casty, C., Wanner, H., Luterbacher, J., Esper, J., and Böhm, R. (2005). Temperature and precipitation variability in the European Alps since 1500. *International Journal of Climatology*, 25:1855–1880; doi:10.1002/joc.1216.

Edwards, T. L., Crucifix, M., and Harrison, S. P. (2007). Using the past to constrain the future: how the palaeorecord can improve estimates of global warming. *Progress in Physical Geography*, 31:481–500; doi:10.1177/0309133307083295.

Esper, J., Büntgen, U., Frank, D. C., Niervergelt, D., and Liebhold, A. (2007). 1200 years of regular outbreaks in alpine insects. *Proceedings of the Royal Society B*, 274:671–679; doi:10.1098/rspb.2006.0191.

Esper, J., Frank, D. C., Wilson, R. J. S., and Briffa, K. R. (2005). Effect of scaling and regression on reconstructed temperature amplitude for the past millennium. *Geophysical Research Letters*, 32(7):L07711; doi:10.1029/2004GL021236.

Handa, I. T., Korner, C., and Hattenschwiler, S. (2005). A test of the tree-line carbon limitation hypothesis by in situ CO_2 enrichment and defoliation. *Ecology*, 86(5):1288–1300; doi:10.1890/04-0711.

Hughes, M. K. (2002). Dendrochronology in climatology - the state of the art. *Dendrochronologia*, 20(1-2):96–116; doi:10.1078/1125-7865-00011.

Hughes, M. K. and Diaz, H. F. (1994). Was there a "Medieval Warm Period", and if so, where and when? *Climatic Change*, 26(2-3):109–142; doi:10.1007/BF01092410.

IPCC (2007). *Climate Change 2007: the Physical Science Basis. Contribution of Working Group I to the Fourth Assessment Report of the Intergovernmental Panel on Climate Change*. Cambridge University Press, Cambridge, United Kingdom and New York, NY, USA.

Jones, P. D., Briffa, K. R., Osborn, T. J., Lough, J. M., van Ommen, T. D., Vinther, B. M., Luterbacher, J., Wahl, E. R., Zwiers, F. W., Mann, M. E., Schmidt, G. A., Ammann, C. M., Buckley, B. M., Cobb, K. M., Esper, J., Goosse, H., Graham, N., Jansen, E., Kiefer, T., Kull, C., Küttel, M., Mosley-Thompson, E., Overpeck, J. T., Riedwyl, N., Schulz, M., Tudhope, A., Villalba, R., Wanner, H., Wolff, E., and Xoplaki, E. (2009). High-resolution palaeoclimatology of the last millennium: a review of current status and future prospects. *The Holocene*, 19(1):3–49; doi:10.1177/0959683608098952.

Kundzewicz, Z., Mata, L., Arnell, N., Döll, P., Kabat, P., Jiminéz, B., Miller, K., Oki, T., Sen, Z., and Shiklomanov, I. (2007). Freshwater resources and their management. In Parry, M., Canziani, O., Palutikof, J., van der Linden, P. J., and Hanson, C., editors, *Climate Change 2007: Impacts, Adaptation and Vulnerybility. Contribution of Working Group II to the Fourth Assessment Report of the Intergovernmental Panel on Climate Change*, pages 173–210. Cambridge University Press, Cambridge, United Kingdom and New York, NY, USA.

Larocque, I., Grosjean, M., Heiri, O., Bigler, C., and Blass, A. (2009). Comparison between chironomid-inferred July temperatures and meteorological data AD 1850-2001 from varved Lake Silvaplana, Switzerland. *Journal of Paleolimnology*, 41(2):329–342; doi:10.1007/s10933–008–9228–0.

Lee, T., Zwiers, F., and Tsao, M. (2008). Evaluation of proxy-based millennial reconstruction methods. *Climate Dynamics*, 31(2):263–281; doi:10.1007/s00382–007–0351–9.

Mann, M. E., Rutherford, S., Wahl, E., and Ammann, C. (2005). Testing the fidelity of methods used in proxy-based reconstructions of past climate. *Journal of Climate*, 18(20):4097–4107; doi:10.1175/JCLI3564.1.

McCarroll, D., Gagen, M. H., Loader, N. J., Robertson, I., Anchukaitis, K. J., Los, S., Young, G. H. F., Jalkanen, R., Kirchhefer, A., and Waterhouse, J. S. (2009). Correction of tree ring stable carbon isotope chronologies for changes in the carbon dioxide content of the atmosphere. *Geochimica et Cosmochimica Acta*, 73(6):1539–1547; doi:10.1016/j.gca.2008.11.041.

McCarroll, D. and Loader, N. J. (2004). Stable isotopes in tree rings. *Quaternary Science Reviews*, 23(7-8):771–801; doi:10.1016/j.quascirev.2003.06.017.

Millennium (2009). Third periodic activity report. In McCarroll, D., editor, *European climate of the last millennium (SUSTDEV-2004-3.1.4.1)*. University of Wales Swansea.

Saurer, M., Cherubini, P., Ammann, M., De Cinti, B., and Siegwolf, R. (2004). First detection of nitrogen from NOx in tree rings: a $^{15}N/^{14}N$ study near a motorway. *Atmospheric Environment*, 38(18):2779–2787; doi:10.1016/j.atmosenv.2004.02.037.

Saurer, M., Maurer, S., Matyssek, R., Landolt, W., Günthardt-Goerg, M. S., and

Siegenthaler, U. (1995). The influence of ozone and nutrition on δ^{13}C in *Betula pendula*. *Oecologia*, 103(4):397–406; doi:10.1007/BF00328677.

Saurer, M. and Siegwolf, R. (2004). Pyrolysis techniques for oxygen isotope analysis of cellulose. In *Handbook of Stable Isotope Analytical Techniques*, volume 1, pages 497–508; doi:10.1016/B978-044451114-0/50025-9. Elsevier, New York.

Sigl, M., Gabriele, J., Bolius, D., Barbante, C., Boutron, C., Gäggeler, H., and Schwikowski, M. (in prep.). High-resolution dust record over the last 1,000 years from an Alpine ice core (Colle Gnifetti, 4450 m a.s.l., Swiss/Italien Alps).

Trachsel, M., Eggenberger, U., Grosjean, M., Blass, A., and Sturm, M. (2008). Mineralogy-based quantitative precipitation and temperature reconstructions from annually laminated lake sediments (Swiss Alps) since AD 1580. *Geophysical Research Letters*, 35(13):L13707; doi:10.1029/2008GL034121.

Treydte, K., Frank, D., Esper, J., Andreu, L., Bednarz, Z., Berninger, F., Boettger, T., D'Alessandro, C. M., Etien, N., Filot, M., Grabner, M., Guillemin, M. T., Gutierrez, E., Haupt, M., Helle, G., Hilasvuori, E., Jungner, H., Kalela-Brundin, M., Krapiec, M., Leuenberger, M., Loader, N. J., Masson-Delmotte, V., Pazdur, A., Pawelczyk, S., Pierre, M., Planells, O., Pukiene, R., Reynolds-Henne, C. E., Rinne, K. T., Saracino, A., Saurer, M., Sonninen, E., Stievenard, M., Switsur, V. R., Szczepanek, M., Szychowska-Krapiec, E., Todaro, L., Waterhouse, J. S., Weigl, M., and Schleser, G. H. (2007). Signal strength and climate calibration of a European tree-ring isotope network. *Geophysical Research Letters*, 34(24):L24302; doi:10.1029/2007GL031106.

von Storch, H., Zorita, E., Jones, J. M., Dimitriev, Y., Gonzalez-Rouco, F., and Tett, S. F. B. (2004). Reconstructing past climate from noisy data. *Science*, 306(5696):679–682; doi:10.1126/science.1096109.

List of Tables

1.1 Records of Northern Hemisphere temperature 8

3.1 Earlywood and latewood $\delta^{13}C$ values 60
3.2 Intercomparison statistics of carbon isotopes in earlywood and latewood 62
3.3 Climate correlations of earlywood and latewood 62

4.1 Site characteristics and detected larch budmoth events 78
4.2 Long-term chronology of detected larch budmoth events 79
4.3 Correlation matrices between larch and spruce for carbon and oxygen isotopes 87

5.1 Description of the Lötschental site 108
5.2 Calibration and verification statistics of $\delta^{13}C$ against the July-August drought index 123
5.3 Central European hydro-climatic series 125
5.4 Correlation matrix between drought index and three datasets of scPDSI 131

6.1 Overview of the different cohorts of trees within the chronology and their origin 148
6.2 Statistical characteristics of carbon and oxygen isotopic composition in the different cohorts of the chronology 153
6.3 Inter-laboratory comparison of alpha-cellulose of standardized wood material 154

I

List of Figures

1.1	Projected surface temperature changes for the 21st century	3
1.2	Records of Northern Hemisphere temperature variation during the last 1,300 years	7
1.3	Current limitations of European palaeoclimate	12
1.4	Temperature predictions with and without palaeoclimatic constraints .	15
2.1	Schematic diagram of carbon isotope variations in trees	35
2.2	Schematic diagram of oxygen isotope variations in trees	38
3.1	Location of sampling sites of the earlywood and latewood study	57
3.2	Intercomparison of carbon isotopes in earlywood and latewood	61
4.1	Temperature correlations of carbon and oxygen isotopes	83
4.2	Wilcoxon-Mann-Whitney rank sum test of budmoth events	85
4.3	Superposed epoch analysis of outbreak patterns in tree-ring data	86
4.4	Intercomparison of stable isotopes in non-host spruce and host-larch . .	88
4.5	Long-term temperature patterns of budmoth outbreak events	89
5.1	The sampling site Lötschental	107
5.2	Carbon chronology characteristics	113
5.3	Oxygen chronology characteristics	114
5.4	Correlation analysis of $\delta^{13}C$ and climate variables	116
5.5	Correlation analysis of $\delta^{18}O$ and climate variables	117
5.6	Significant spatial field correlations between the two isotope series and temperature and precipitation	119

List of Figures

5.7 Comparison of Lötschental δ^{13}C with climate reconstructions 120
5.8 Multi-taper method spectra of δ^{13}C, temperature, precipitation and the drought index 122
5.9 Carbon-isotope series and drought index: raw and high/low pass filtered 124
5.10 δ^{13}C drought reconstruction (LOT$_{DRI}$) and comparisons 126

6.1 The sampling sites Lötschental and Simplon including living trees and historical buildings 145
6.2 Tree-line sites in the Lötschental 149
6.3 "Gemeindestadel" in Blatten, Lötschental 149
6.4 "Stallscheune" in Eisten-Gryn, Lötschental 149
6.5 "Stall Dorsaz/Guntern" in Simplon Dorf, Simplon 149
6.6 Annually resolved carbon and oxygen raw chronologies for AD 800-2004 152
6.7 Carbon and oxygen isotopes in the first overlap period 154
6.8 Carbon and oxygen chronologies and their overlap periods after standardization of each segment 156
6.9 Multi-taper method spectra of δ^{13}C and δ^{18}O 157
6.10 Carbon and oxygen chronologies after applying a 30-year low pass filter 159
6.11 AD 800–2004 extreme events in various tree-ring parameters 160

Acknowledgements

This thesis would not have been possible without the manifold support of numerous people especially from the three institutions PSI, ETH and WSL. I would like to thank all of them, realizing that those whom I owe most I cannot thank enough, and that things for which I am most grateful, I cannot put adequately into words.

First of all I would like to thank Harald Bugmann for accepting me as his PhD student, for his engaged supervision, for stimulating discussions and for the careful review of our manuscripts. For the opportunity to conduct this thesis in his group I am especially thankful to Rolf Siegwolf, and I strongly appreciated his generous support, his enthusiasm, and his endless flow of ideas. Many thanks to Matthias Saurer, for the supervision of this thesis. Although, at times, our approach to things can be very different, I can't thank him adequately for his careful guidance, and for the freedom to explore different research approaches; without his support in the lab, his deep insight into the world of stable isotopes, his ideas, comments and suggestions, this thesis would have never become what it is. I sincerely thank Neil Loader (University of Wales Swansea) for his spontaneous and immediate positive response to fit this co-examination into his tight schedule.

Furthermore, I am very thankful to Urs Baltensperger for accepting me as a PhD student in his laboratory, for his particular interest in my thesis – although treating rather unfamiliar topics –, for his honest comments and for the introduction into the world of aerosol science. Being a LAC-member, I had a fabulous technical and administrative support by Günther Wehle, René Richter, Bettina Möhrle, Michel Tinguely, Catharina Lötscher and Doris Hirsch-Hoffmann. Thank you all very much. In addition, I would like to thank all "my" practica students, apprentices and vacation jobbers:

Lara Läubli, Melanie Tröndle, Nicolas Bircher, Lukas Bächli, Jonas Schmid and Sandro Lüscher. Your remarkable work and particular interest extended the chronologies to their current length and it was a real pleasure working with all of you.

I also want to take the opportunity to thank the members of the dendro-unit at WSL and of the Forest Ecology group at ETH for making me feel home in their groups. I sincerely thank Jan Esper, David Frank, Kerstin Treydte and Ulf Büntgen for their numerous constructive inputs and discussions that broadened my knowledge in the wide field of dendrochronology and life as a scientist. Many thanks to Anne Verstege, Daniel Nievergelt and Martin Schmidhalter (Dendrolabor Valais) for tremendous support in the field and in the dendro-lab. I also would like to thank Christof Bigler and Patrick Weibel who shared with me their rich experience in drought indices.

I am very grateful that I could conduct my thesis within the framework of the EU-project "Millennium" (FP6-2004-GLOBAL-017008-2). Besides financial support, the "Millennium-family" was an excellent basis for lively discussions and interdisciplinary exchange and enabled joint-work on the chronologies. In this context, thanks to Sarah Hangartner and Markus Leuenberger, who took care of sample preparation and measurements of part of the historic timber from the Lötschental. In particular, I would like to express my sincere thanks to Danny McCarroll; I enjoyed very much his exceptional interest in my work, our discussions about the dominating climate signal in my isotope series and the approaches of dealing with vertical uncertainties. Furthermore, many thanks to Valerie Hall and Sheila Hicks for their extraordinary encouragement and the friendly and inspiring atmosphere during the "Millennium"-paperwriting workshops. Moreover, a big thank you to Giles Young; writing a paper with you was a real pleasure!

For keeping me mentally and physically fit, I thank my running mates and SOLA companions (Agnes, Rahel, Lukas, Peter B., Josef, Daniel, Iakovos, Martin, Francesco, Claudia, Zsofie, Arnaud, Peter M., Maarten, Nolwenn and Paul). Besides picking up the pace, I very much enjoyed our discussions on all aspects of life and I am especially grateful to those who revealed excellent reviewer qualities towards the end of my thesis. I also would like to thank the non-runners, in particular Torsten and Anita, for the thoroughly review of some chapters of my thesis and the motivating and inspiring coffee-breaks.

Special thanks to Axel, Jonathan, Kathrin, Silke, Ingo, Stephan and Christina. Without your friendship I would have been pretty lost in the confusing world of science.

Above all, I want to thank my family and my friends for their manifold support during all my studies. Lastly, and most important thank you to Hendrik for making my life meaningful and rewarding far beyond anything science can explain.

Die VDM Verlagsservicegesellschaft sucht für wissenschaftliche Verlage abgeschlossene und herausragende

Dissertationen, Habilitationen, Diplomarbeiten, Master Theses, Magisterarbeiten usw.

für die kostenlose Publikation als Fachbuch.

Sie verfügen über eine Arbeit, die hohen inhaltlichen und formalen Ansprüchen genügt, und haben Interesse an einer honorarvergüteten Publikation?

Dann senden Sie bitte erste Informationen über sich und Ihre Arbeit per Email an *info@vdm-vsg.de*.

Sie erhalten kurzfristig unser Feedback!

VDM Verlagsservicegesellschaft mbH
Dudweiler Landstr. 99
D - 66123 Saarbrücken
Telefon +49 681 3720 174
Fax +49 681 3720 1749

www.vdm-vsg.de

Die VDM Verlagsservicegesellschaft mbH vertritt

Printed by Books on Demand GmbH, Norderstedt / Germany